Perspectives on the Teaching of Mathematics

Sixty-sixth Yearbook

Perspectives on the Teaching of Mathematics

Sixty-sixth Yearbook

Rheta N. Rubenstein
Sixty-sixth Yearbook Editor
University of Michigan—Dearborn
Dearborn, Michigan

George W. Bright
General Yearbook Editor
University of North
Carolina at Greensboro (Emeritus)
Greensboro, North Carolina

NATIONAL COUNCIL OF
TEACHERS OF MATHEMATICS

1906 Association Drive, Reston, VA 20191-1502
(703) 620-9840; (800) 235-7566; www.nctm.org

Library of Congress Cataloging-in-Publication Data
Perspectives on the teaching of mathematics : sixty-sixth yearbook / [edited by] Rheta N. Rubenstein, George Bright. p. cm.
ISBN 0-87353-558-8 (hardcover)
1. Mathematics--Study and teaching--United States. I. Rubenstein, Rheta Norma Pollock. II. Bright, George W.
QA13.P4725 2004
510'.71'073--dc22

2004002323

The National Council of Teachers of Mathematics is a public voice of mathematics education, providing vision, leadership, and professional development to support teachers in ensuring mathematics learning of the highest quality for all students.

The publications of the National Council of Teachers of Mathematics present a variety of viewpoints. The views expressed or implied in this publication, unless otherwise noted, should not be interpreted as official positions of the Council.

Printed in the United States of America

Contents

Part 2: The Enactment of Teaching

Part 3: Support of Teaching

Preface

Teaching is a complex endeavor. The knowledge base behind mathematics teaching includes the knowledge of mathematics, of connections among mathematical ideas, of students, of students' learning, of school culture, and much more. The process of teaching involves creating a learning community, challenging students to make sense of mathematical ideas, and supporting students' developing understanding. The teaching process involves a myriad of decisions, of which just a few are "When do we tell?" "When do we ask more questions?" and "When do we merely encourage more thinking?" Consequently, it is not surprising that learning to teach well is a career-long endeavor! With roots in preservice education, it is sustained throughout one's career in an ongoing process of learning what students understand, how they understand it, and what learning activities most effectively support meaningful and useful understanding. This yearbook is organized around three aspects of teaching: foundations for teaching, the enactment of teaching, and the support of teaching nurtured in preservice education and strengthened throughout a teacher's career.

Part 1 provides perspectives on a few of the many underlying foundations. Carol Malloy begins by addressing an overarching concern across the entire teaching spectrum: equity—the concept of significant mathematics being learned meaningfully by all students. She shares research and conceptual frameworks that ultimately converge into a vision of excellent teaching, effective for all students. Jennifer M. Bay-Williams, Elizabeth M. Skipper, and Susan K. Eddins argue for the value of understanding entire trajectories of mathematical ideas, from primary through high school, when designing instruction at any level. Curriculum coherence and understanding where students have been and where they are going are essential for effective teaching and learning. Within the frame of curriculum and materials, Fran Arbaugh and Catherine A. Brown ask, "What Makes a Mathematical Task Worthwhile?" If we know that worthwhile tasks are crucial learning tools, then what are their characteristics? Having criteria by which to recognize or create such tasks is essential in teachers' work. Arbaugh and Brown offer suggestions. M. Kathleen Heid, Rose Mary Zbiek, and Glendon W. Blume note that when technology plays a role in instruction, teachers' mathematical foundations may need to be deepened. These authors make specific suggestions for the study of functions. Finally, Carne (Barnett) Clarke and Alma Ramirez highlight the difficulties inherent in students' use of language: students' construction of meanings and their use of language are tightly interwoven. Clarke and Ramirez identify potential pitfalls and suggest pathways around them.

Part 2 addresses the enactment of teaching. Along with colleagues, Barbara and David Clarke have done extensive research on primary school students' learning. Where learning was exceptionally strong, they and fellow researchers looked closely at the children's teachers. As a result they have been able to identify several characteristics associated with effective teaching. Also at the primary school level, Aki Murata, Naoyuki Otani, Nobuaki Hattori, and Karen C. Fuson provide a detailed picture of a Japanese classroom taught in the United States. Through this example they illustrate how teachers communicate their valuing of all students' responses while at the same time moving students toward specific instructional goals. Margaret Smith, in a similar vein, draws from the video data of the Third International Mathematics and Science Study (TIMSS) and contrasts elements of teaching algebra lessons typical of the United States and Japan. Debra I. Johanning and Teri Keusch provide images from a specific middle grades fractions lesson that suggest how classroom learning communities can be built. Moving beyond classrooms, Phyllis Whitin suggests extending learning communities into homes. She gives specific suggestions and guidelines for involving family members in thinking together with children about intriguing mathematical ideas.

Although problem solving has been a centerpiece of mathematics instruction for many decades, the role of problem posing has gotten less attention. David J. Whitin helps remedy this situation by illustrating cases of problem posing at many instructional levels and by highlighting the instructional benefits of encouraging students to pose, as well as solve, problems. Joan Cohen Jones's article closes Part 2 by revisiting the issue of equity with an array of ideas drawn from a variety of classrooms and sources.

Teachers think regularly about supporting students' learning. Less often do teachers or the public think about supporting teachers' learning. Part 3 addresses this crucial element in building effective teaching systems: the career-long support of teaching. The first three articles address teachers' content knowledge. Charlene E. Beckmann, Pamela J. Wells, John Gabrosek, Esther M. H. Billings, Edward F. Aboufadel, Phyllis Curtiss, William Dickinson, David Austin, and Alverna Champion represent a team of mathematicians and mathematics educators who have worked together to strengthen mathematics courses taken by many undergraduates including future teachers. Their approach has involved adapting and extending *Standards*-based grades K–12 curricular activities to the college classroom. They provide examples from several courses where the Reasoning strand is highlighted. At the secondary school level, Irene Bloom shares how challenging problems advance prospective teachers' mathematical understanding as they make connections within and across mathematics strands. Similarly, Kathleen Cramer engages in-service elementary school teachers through an array of problem-solving activities focused on big ideas. Her work reveals factors that reduce anxiety and build confidence in doing mathematics.

The next several articles concentrate on knowledge where teaching strategies and mathematical content knowledge intersect. Theresa J. Grant and Kate Kline have done a remarkable job of building trust among in-service elementary school teachers learning to use reform curricular materials. Participating teachers willingly allow themselves to be videotaped and to have colleagues observe and analyze their lessons. Together, colleagues learn to deal sensitively and thoughtfully with many vital decisions in teaching. Lynn C. Hart, Deborah Najee-ullah, and Karen Schultz, too, have built trusting relationships with teachers, in part by opening their own work to others through modelling a number of strategies including critical self-reflections. As a result, teachers are more willing to analyze objectively their own teaching and that of colleagues. P. Mark Taylor's research indicates that when teachers work together within schools, features of administrative support and the culture of the school may support or hinder their collaboration. Related to issues of school culture, the article by Patricia A. Jaberg, Cheryl A. Lubinski, and Sigrid Aeschleman highlights aspects of the principal's role in giving support for teachers' professional development.

Thinking carefully about assessment as another avenue for professional development is illustrated in the article by Sandra Wilcox and Elizabeth M. Jones. A sample problem from the Balanced Assessment project is shared with multiple implications for teachers' reflection and for students' activities. Professional development related to technology poses its own challenges. Karen Hollebrands and Rose Mary Zbiek furnish a road map for the stages of professional growth related to the integration of technology into teaching. Finally, Janet Warfield and Cheryl Lubinski delve into the realm of supporting teachers as they begin to assume leadership roles with colleagues. Their work poses a crucial question, "If we believe students learn well by struggling with challenges, then how do we extend that philosophy to our work with new teacher leaders?"

Throughout, authors have shared beliefs, insights, experiences, and findings that inform, intrigue, and expand our perspectives of mathematics teaching. As I look back at the articles and the Teaching Principle of the *Principles and Standards for School Mathematics* (National Council of Teachers of Mathematics [NCTM] 2000), the original impetus for this yearbook, I hear many of the messages of that Principle echoed in the authors' voices.

"To be effective, teachers must know and understand deeply the mathematics they are teaching" (NCTM 2000, p. 17) is the foundation for the articles by Beckmann and colleagues; Bloom; Cramer; and Heid, Zbiek, and Blume.

“Effective mathematics teaching requires a serious commitment to the development of students’ understanding of mathematics.... Teaching mathematics well involves creating, enriching, maintaining, and adapting instruction to move toward mathematical goals, capture and sustain interest, and engage students in building mathematical understanding” (NCTM 2000, p. 18). For more details, see the articles by Barbara and David Clarke; Murata and colleagues; Phyllis Whitin; and Bay-Williams, Skipper, and Eddins.

“Effective teaching requires a challenging and supportive classroom learning environment.... Effective teaching conveys a belief that each student can and is expected to understand mathematics and that each will be supported in his or her efforts to accomplish this goal” (NCTM 2000, p. 18). This statement is the heart behind the articles by Malloy and by Joan Cohen Jones.

“Are students’ discussion and collaboration encouraged? Are students expected to justify their thinking? If students are to learn to make conjectures, experiment with various approaches to solving problems, construct mathematical arguments and respond to others’ arguments, then creating an environment that fosters these kinds of activities is essential” (NCTM 2000, p. 18). Scenes with these features are found in the articles by Johanning and by Smith.

“In effective teaching, worthwhile mathematical tasks are used to introduce important mathematical ideas and to engage and challenge students intellectually. Well-chosen tasks can pique students’ curiosity and draw them into mathematics” (NCTM 2000, p. 18). Arbaugh and Brown elaborate on what constitutes a worthwhile mathematical task.

“Effective teaching involves observing students, listening carefully to their ideas and explanations, ... and using the information to make instructional decisions” (NCTM 2000, p. 19). These and related issues of communication are discussed by Clarke and Ramirez.

“Opportunities to reflect on and refine instructional practice—during class and outside class, alone and with others—are crucial in the vision of school mathematics outlined in *Principles and Standards.* To improve their mathematics instruction, teachers must be able to analyze what they and their students are doing and consider how those actions are affecting students’ learning” (NCTM 2000, p. 19). Reflection is a major component in the work of Grant and Kline and that of Hart, Najee-ullah, and Schultz.

“Collaborating with colleagues regularly to observe, analyze, and discuss teaching and students’ thinking ... is a powerful, yet neglected, form of professional development in American schools (Stigler and Hiebert 1999)” (NCTM 2000, p. 19). Hollebrands and Zbiek, Warfield and Lubinski, and Taylor address collaboration.

“Using a variety of strategies, teachers should monitor students’ capacity and inclination to analyze situations, frame and solve problems, and make sense of mathematical concepts and procedures. They can use this information to assess their students’ progress and to appraise how well the mathematical tasks, student discourse, and classroom environment are interacting to foster students' learning" (NCTM 2000, p. 19). See David Whitin for having students "frame" problems and Wilcox for the uses of assessment to help make instructional decisions.

"The work and time of teachers must be structured to allow and support professional development that will benefit them and their students" (NCTM 2000, p. 19). The principal's role, one aspect of structuring the work of teachers, is discussed by Jaberg, Lubinski, and Aeschleman.

Accompanying this yearbook is a companion professional development handbook. The handbook offers teachers and teacher educators guidance in bring-ing to life issues and ideas from this yearbook in teacher education activities.

The work of creating this book was succored and guided by the thoughtful and enthusiastic work of a hard-working and caring panel:

George Bright, *University of North Carolina at Greensboro (retired), Greensboro, North Carolina*

Morton Brown, *University of Michigan—Ann Arbor (retired), Ann Arbor, Michigan*

Betty Causey-Lee, *Detroit Public Schools, Detroit, Michigan*

Wendy Rich, *Randolph County Schools, Asheboro, North Carolina*

Edna Vasquez, *Southfield High School, Southfield, Michigan*

George Bright, the general editor for the three NCTM yearbooks published in 2002–2004, was a special source of insight and encouragement.

I would like to thank heartily all the authors who submitted manuscripts for the original call. We could not publish everything received, but I know that many thoughtful colleagues have wonderful ideas to share. Their willingness to write and submit their work is part of what makes our profession grow. To the authors whose work appears here, thanks for your efforts and your patience in the many rounds of writing and revising. To the Reston staff, including Charles Clements, David Webb, Glenn Fink, and Nick Abrash, I am honored to have had the opportunity to work with such a capable and considerate group.

Rheta N. Rubenstein
Sixty-sixth Yearbook Editor

REFERENCES

National Council of Teachers of Mathematics (NCTM). *Principles and Standards for School Mathematics.* Reston, Va.: NCTM, 2000.

Stigler, James W., and James Hiebert. *The Teaching Gap: Best Ideas from the World's Teachers for Improving Education in the Classroom.* New York: Free Press, 1999. Quoted in National Council of Teachers of Mathematics (NCTM), *Principles and Standards for School Mathematics* (Reston, Va.: NCTM, 2000), p. 19.

1

Equity in Mathematics Education Is about Access

Carol E. Malloy

CALLS for equity in mathematics education are well known. What is less known is what is needed to achieve this important goal. Currently mathematics classrooms include students from diverse cultures, ethnic groups, socioeconomic levels, and achievement levels, as well as other students with limited opportunities to participate fully in mathematics. Within diverse student populations, African American, Native American, Hispanic, and poor students are more likely to have fewer opportunities to participate in higher-level mathematics classes; thus their mathematics achievement does not meet parity with the total population. Educators have offered numerous reasons for the lagging achievement including a discontinuity of home and school culture, students' dissociation with academic achievement, tracking, instruction in opposition to students' learning preferences, and a disparity in resources and trained teachers in schools with large minority populations (Gay 2000; Kozol 1991; Stiff 1990; Steele 1992). Regardless of the reason, a large number of students do not have the opportunity to learn. This article reviews research and provides examples to clarify what can be done so that all children achieve strong mathematics learning.

When mathematics educators discuss equity in mathematics teaching and learning, they refer to *mathematics for all children* or *mathematics opportunity for every child.* Inherent in this concept is the desire to furnish every child the high-quality mathematics education that will give them access to professions and careers of their choice. Equity in mathematics education is about access. However, we have not reached measurable equity in achievement. This means that students do not have the opportunity to acquire the skills needed to access 70 percent of the careers of today (Moses and Cobb 2001). Schoenfeld (2002), in discussing the "potential for providing high quality mathematics instruction for all students" (p. 13), identifies four systemic conditions needed for achieving this goal: "(*a*) high quality curriculum; (*b*) stable, knowledgeable, and professional teaching community; (*c*) high quality assessment that is aligned with curricular goals; and (*d*) stability and mechanisms for the evolution of curricula, assessment, and professional development" (p. 13). These conditions are appropriate and necessary for schools and districts to consider and implement as they progress toward the important goal of equity and access for all students in mathematics. They involve policy and efforts well beyond classrooms. This article focuses on just one part of the solution: classroom culture and instruction. Because there is no one way to address this goal, I take a descriptive approach rather than a prescriptive approach. First, I present a framework through which teachers can make instructional decisions that are pivotal in creating classrooms where students are successful. This framework focuses on giving students the opportunity to acquire conceptual understanding of the mathematics. Second, I discuss the learning preferences of students. Finally, I relate those preferences to the vision of mathematics instruction described in the National Council of Teachers of Mathematics (NCTM) *Standards* documents (NCTM 1989, 1991, 2000).

FRAMEWORK FOR INSTRUCTIONAL DECISIONS

The reform of mathematics instruction envisioned in the NCTM *Standards* documents requires a shift in how educators think about teaching and learning mathematics. There are three facets to the reform: *curriculum* that gives students access to valuable mathematics, *pedagogy* that allows students to become doers of mathematics, and *assessment* that guides the practice of teachers. Central to these facets are the role of the teacher, the learning culture that teachers establish, and how students interpret that culture. Teachers who have a serious commitment to developing students' conceptual understanding have equitable expectations and practices, use pedagogically appropriate instructional practices, and use curriculum materials and assessment methods that promote students' conceptual development. They give students problem-solving opportunities using valuable and meaningful mathematical tasks where students learn mathematics through a sociocultural approach that requires that students share, question, challenge, and explore mathematics together (Grouws and Lembke 1996). Teachers also provide a learning environment that is nonthreatening and where students' ideas are valued (Sowder and Phillipp 1999). Because teachers and students have varied styles of teaching and learning, it is important to recognize that there is no "one" correct way to teach mathematics. However, the experiences teachers offer have a direct impact on students' understanding of mathematics, their abilities to solve problems, and their confidence in and disposition toward their personal mathematics learning (NCTM 2000).

The following is a framework for establishing classrooms that are equitable and provide access through conceptual learning of mathematics. The framework contains an approach to examine students' development of mathematical understanding, the roles and decisions of teachers as they foster the development of conceptual understanding, and a structure through which to promote social interactions in the classroom. The combination of the three components offers a means for teachers to identify, create, and maintain a learning culture in classrooms that is conducive to students' gaining conceptual understanding of mathematics.

Conceptual Understanding

Carpenter and Lehrer (1999) identified five forms of mental activity that help students gain mathematical understanding. They characterize understanding not as a static attribute but as emerging in learners with the following interrelated forms: (*a*) constructing relationships, (*b*) extending and applying mathematical knowledge, (*c*) reflecting about experiences, (*d*) articulating what one knows, and (*e*) making mathematical knowledge one's own (p. 20). Using the Carpenter and Lehrer model, students construct relationships from prior knowledge and experiences. They construct meanings of new mathematical ideas by relating the new ideas to knowledge they already have. Carpenter and Lehrer explain that developing understanding requires more than connecting new and prior knowledge; it requires a structuring of knowledge so that new knowledge can be "related to and incorporated into existing networks of knowledge rather than connected on an element-by-element basis" (p. 21). Students are reflective about learning when they are aware of the knowledge they have acquired and examine it in relationship to what they know. This metacognitive process of reflection leads to a reorganization of the students' mathematical knowledge. Another step in students' understanding, articulating what they have learned, is a public form of reflection. Finally, this five-tiered process of learning with understanding is accomplished with students involved in activities that allow them to be personally involved in the development of knowledge, resulting in students' making the knowledge they have developed their own. Clearly not all students progress through these mental activities at the same level. However, how and if students progress through these five mental activities should provide some indication of the depth of understanding students acquire.

Instructional Decisions of Teachers

In conjunction with this generative process of students' understanding mathematics, the teacher must make instructional decisions about the (*a*) content being taught, (*b*) tasks students are asked to complete, (*c*) pedagogy appropriate for the lesson, and (*d*) assessment the teacher uses to determine what students have learned (Pilburn et al. 2000; Shafer 2001).

The content that a teacher selects is based on curriculum guides provided by the state or district, but teachers decide how the content is presented. For instance, in teaching students to find a percent represented by a fraction, a teacher could plan to show students an algorithm or help them figure out what a percent means, or both. Related to these decisions is the teacher's understanding of the fundamental mathematical ideas in the content. Associated with content are tasks that students are asked to perform, an important medium through which the mathematics is learned. Looking at tasks can help determine if teachers are expecting students to experience mathematics with high-level cognitive demands or low-level cognitive demands (Stein et al. 2000). Using the percent example, students who are only taught to find the percent represented by 1/3 by dividing one by three and writing the decimal equivalent as a percent using an algorithm are presented with a task with low-level cognitive demands. This method often results in only procedural understanding. However, if students are asked first to estimate the percent by recognizing that there are three 1/3's in a whole and that a whole is 100 percent, they can use trial and error or other methods to estimate 1/3 as about 33 percent. Then they can find the exact percent by realizing that 3×33 is 1 percent less than 100 percent; and since three 1/3's equal 1, they will see there will be 1/3 more associated with the percent. Thus they know that 1/3 = 33 1/3 percent. This task has high-level cognitive demands and promotes conceptual understanding of equivalent representations of fractions and percents. A combination of both approaches gives students an understanding of percents and a method to quickly find the percent associated with a fraction.

Pedagogy and assessment are intertwined. Pedagogical decisions are based on methods of instruction, sequencing, timing, and the complexity of mathematical tasks (Shafer 2001). Decisions involve teachers' determining where students might have difficulty with new topics, teachers' knowledge of students' typical learning sequences, and their ability to diagnose students' thinking and identify methods to help students overcome difficulties (Shafer 2001). Classroom assessment includes how teachers listen to students' conversations and responses and how they review students' written work. Assessment, both formative and summative, includes the evidence a teacher uses to determine students' understanding, the feedback the teacher gives to students about their work, and the reviews of students' critiques of their work.

Mathematical Interaction

Interaction in mathematics instruction relates to classroom discourse, types of problems posed to students, the level of questioning offered by the teacher and students, the reciprocity of knowledge exemplified by students' conversations, and how they listen to one another. Looking at how students challenge and justify solutions and explanations can enhance classroom communication as students solve problems, explain their understanding and interpretations, and validate their work (Cobb et al. 1992). Interactions are also seen in Carpenter and Lehrer's (1999) five forms of mental activity that help students gain mathematical understanding: constructing relationships, extending knowledge, reflecting about experiences, articulating what one knows, and making mathematical knowledge one's own.

This framework for establishing equitable classrooms considers the development of students' conceptual understanding, instructional decisions of teachers, and mathematical interactions in the classroom. Below is a narrative describing a first-year teacher's classroom where students experienced an effective and equitable culture for learning mathematics.

A Classroom Example

In October 2002, I observed a sixth-grade classroom where twenty-one racially diverse students were learning to interpret graphs that involve how a person becomes more skillful at a task. The teacher taught mathematics and science in the same room; therefore, the students were seated at large science tables in horizontal rows across the classroom. The class began with a warm-up problem that students completed while the teacher walked around the room checking students' homework. Before the teacher began instruction, she asked the students to explain what they had done the previous day. Three students eagerly volunteered that they had learned how to read line graphs; how to graph data using a line graph; and how to see that, as a person's skill increases, the time it takes to complete a task gets shorter. Today the teacher asked the students to investigate a new situation. Students were to interpret data about a girl who was learning to skate on Roller blades, or inline skates. In this exercise data represented the number of times the girl fell in her first eight one-hour skating periods. The teacher's goal was for students to understand that the context of a problem is important to the interpretation of the data. As the skill level of the girl increased, the number of times that she fell decreased.

Working as a whole class, the students were orderly and seemed enthusiastic to complete the assignment. The teacher questioned students as they found the mean, median, mode, and range of the data and made a line graph of the data. If students gave a wrong answer, other students often challenged them. In these situations, the teacher did not say, "You are wrong." Instead, she asked students to rethink answers and gave students time to think and redo exercises. As a result, students were willing to take risks and volunteer because they knew they would learn even if they were not correct. This teacher asked questions and pursued students' questions that allowed for divergence and discussion rather than just convergence toward the content to be learned. For instance, when the teacher asked the students what conclusions could be made from the number of falls decreasing from 8 to 3 falls over the eight sessions, students responded that the longer she practiced, the better she got at inline skating. One student questioned the data and asked, "What if the number of falls did not decrease?" The teacher redirected the question to the class. Students then discussed differences that could have occurred in the terrain where she skated, the weather, and that the crowds in the area could all be factors in how she progressed as an inline skater. Even though the teacher did not plan this discussion, she allowed for students to discuss the possibilities. As a result, the students understood that the conditions surrounding the inline skating influenced their conclusions about the skill level the girl had achieved.

Near the end of the class period, the teacher had students independently write a letter that described how to read and interpret their line graph. She walked around the room helping students by answering their questions. Then several students read their letters aloud. Afterward, through the teacher's questions, students explained what the data showed and how this problem was different from the work they had done previously. To have the students understand that the context of the problem is necessary in the interpretation of data, she ended the class by asking them what would occur in a situation where a student was practicing free throws in basketball. Students understood that in this instance the number of successful free throws would increase as the skill level increased.

As I reflected on the lesson and the follow-up conversation with the teacher, it was clear that this teacher was cognizant of important elements that produce learning experiences that are both effective and equitable. The teacher gave her students the opportunity to gain conceptual understanding by moving them through four of the five forms of mental activity (Carpenter and Lehrer 1999).

1. *Constructing relationships:* Students used their knowledge and experience with skating and line graphs to complete the activity.
2. *Extending relationships:* Students had to think about the number of times the girl fell rather than the time it took for her to complete the task. They also learned that data values could increase or decrease depending on the context.
3. *Reflecting about experiences:* Through discussion the teacher had students think about the activity to formulate a conjecture about the times the girl fell in relationship to her skill level.
4. *Articulating what they know:* Students had to write and read aloud what they understood about the interpretation of data. They had to synthesize what they had learned and determine the important ideas of the activity. Even though in this class period there was no opportunity

to see student ownership of the knowledge gained, many of the students appeared to be excited about the importance of context in the interpretation of data.

This classroom exemplified a classroom where all students had access to high-quality mathematical knowledge. In her role, the teacher accomplished the content, task, pedagogy, and assessment portions of the framework presented above. She also engaged all students in the discussions about the mathematics presented, and all students seemed to be learning. Students were given opportunities to learn important mathematical concepts and procedures with understanding. Students were confidently engaged in complex mathematical tasks that drew on their prior knowledge. They used data to make conjectures about the girl's skating skills, and they supported their conclusions. The teacher guided the students through the learning process with superior questioning techniques, being careful not to tell answers or to do the mathematics for them. She allowed students to answer questions of other students and pursued divergent questions of students. At the end of the lesson students wrote about what they had done and confidently recited what they had learned to the class. This teacher had created a community of learners where students were actively engaged in learning mathematics, and as a result they valued the knowledge that they gained. These are important images of a classroom that embrace the vision of *Principles and Standards for School Mathematics* (NCTM 2000).

Learning Preferences

As mentioned earlier, the publication of the NCTM *Standards* documents established a new vision for school mathematics education. This vision is grounded in equity and knowledge of students' learning with a goal of helping educators become proactive in accommodating instructional and developmental needs of students. The goals align mathematics instruction with the learning processes, preferences, and development of students. Another set of literature, learning-preference research from the late twentieth century, also has as its goal the alignment of instruction to learning processes. Consider the relationships among the researches of Hale-Benson (1986), Hilliard (1976), Shade (1989), and Willis (1992), which suggest that there are specific ways that the majority of African Americans approach learning, accommodations to their findings, and the vision and goals of the NCTM *Standards* documents. Although references here address the preferences of African American students, these preferences in many instances dominate the learning of students from many minority, gender, and socioeconomic groups that are considered outside the mainstream.

Learning Preferences and Mathematics Learning

Research shows that students from different ethnic, cultural, and economic groups—students whose achievement, as groups, is below that of the total population—often approach learning differently. There are preferences for students' interaction with the environment, and these preferences influence cognition, attitude, behavior, and personality. They may be different from those used by the majority of students, but they are not deficient. (For more information, see Malloy 1997; Malloy and Malloy 1998).

Traditionally, mathematics instruction has addressed the needs of the analytic, field-independent, individual learner. Students were instructed in ways that encouraged them to focus on detail and use sequential or structured thinking, recall abstract ideas and irrelevant detail, engage in inanimate material, respond to intrinsic motivation, focus on the task, learn from formal lecture, achieve individually, and emphasize facts and principles. However, many students learn in ways characterized by factors of social or affective emphasis, harmony with their community, holistic perspectives, field dependence, expressive creativity, and nonverbal communication (Stiff 1990; Willis 1992). Their use of verbal expression and holistic approaches incorporate elaborate verbal and motor skills to communicate mathematical ideas and relationships. They may use creative stories or verbal elaboration to convey specific mathematical meanings and often use holistic reasoning in mathematical problems where synthesis is an important strategy in finding problem solutions (Malloy 1994). If all students are to be given access to high-quality mathematics instruction, the instruction must reflect their learning preferences.

Learning preferences and approaches of these learners help us identify several areas where teachers can make accommodations. Students should be allowed to focus on the whole, use improvisational intuitive thinking, recall relevant verbal ideas, engage in human and social content material, respond to extrinsic motivation, focus on interests, learn from informal class discussion, achieve interdependently, and narrate human concepts (Dance 1997; Hale-Benson 1986; Hilliard 1976; Shade 1989; Willis 1992).

Accommodations

Historically, educators did not fully examine the role of nonmajority culture on cognition, and thus they did not contextualize their mathematics instruction to these students' learning preferences. Harris (1994) explains that the process requires an accommodation by the teacher. "This does not mean that allegiance to or identification with anyone's individual culture is denied or denigrated. It does mean, however, that a common ground is created wherein interaction can occur that is meaningful to those involved" (p. 78). Contextualization occurs when mathematics educators consider cultural influences on learning and thus restructure their pedagogy and accommodate their students. On the basis of the preferences summarized above, the restructuring of pedagogy for all students includes the teacher's (*a*) acknowledging and using individual students' preferences in acquisition of knowledge, (*b*) developing activities that promote mathematical discourse within the classroom among students and between the students and teacher, (*c*) valuing students' discourse and verbal knowledge, (*d*) creating interdependent learning communities within the classroom, and (*e*) encouraging, supporting, and providing feedback to students as they learn. The accommodation also includes the creation and use of mathematical tasks that require students to "do mathematics," as well as expectations that students can and will achieve conceptual and procedural understanding of the mathematics content from the framework presented in the first section of this article.

The Vision of the NCTM *Standards*

In the *Standards* documents, NCTM proposed a shift away from teaching mathematics where students were passive individuals, the teacher was the sole authority on right answers, mathematics learning was the memorization of procedures and mechanistic answer-finding, and mathematics was a body of isolated concepts and procedures (NCTM 1989). NCTM recommended a move toward conceiving classrooms as mathematical communities employing logic and mathematical evidence as verification for answers; using mathematical reasoning, conjecturing, inventing, and problem solving; and connecting mathematics, its ideas, and its applications (NCTM 1991). The intended outcomes of the shift in mathematics teaching include having students learn to value mathematics, become confident in their own ability, become mathematical problem solvers, and learn to communicate and reason mathematically (NCTM 1989). NCTM's *Principles and Standards for School Mathematics* (2000) supports the need for a shift in mathematics teaching and learning by grounding the document in the "belief that all students should learn important mathematical concepts and processes with understanding" (p. ix). *Principles and Standards* provided six Principles that "describe particular features of a high-quality mathematics program" (p. 11). The first of these Principles, the Equity Principle, states, "Excellence in mathematics education requires equity—high expectations and strong support for students" (p. 12). The three sections of the principle discuss how equity requires (*a*) high expectations and worthwhile opportunities for all, (*b*) accommodating differences to help everyone learn mathematics, and (*c*) resources and support for all classrooms and all students.

Comparison

Figure 1.1 uses learning preferences of African American and other students as the context, teacher accommodations as the process, and NCTM's vision of school mathematics as the outcome. There are clear connections in every category among students' learning preferences, accommodations for students' learning, and outcomes recommended for students by the NCTM *Standards* documents. Thus, the realization of mathematics instruction recommended by the learning-preference literature, teacher accommodations, and NCTM *Standards* documents concur. Together they offer new learning opportunities for students who have had learning conflicts with the formalized structure of traditional classrooms and may offer continued success and expanded knowledge for students who have always achieved. Recent reports and research support these perspectives. Published studies and official reports indicate that when these Principles are effectively applied, progress has been made in the mathematics achievement of both minority and majority students. Reports of the Longitudinal Evaluation of School Change and Performance in Title 1 Schools, National Science Foundation (NSF) Urban Systemic Districts, NSF Ohio State Systemic Initiative Project Discovery, and NSF Local Systemic Change in North Carolina indicate gains in enrollment and performance by minority students when *Standards*-based instruction and materials were used (Jonsson 2001; Jane Kahle, personal communication, May 21, 2001; National Science Foundation 2001;U.S. Department of Education 2001). Studies completed by Briars and Resnick (2000) found that African American students in strong-implementation *Standards*-based curriculum schools outperformed both white and African American students in weak-implementation schools on the New Standards Mathematics Reference Exam. Reys et al. (2003) found that middle grades students, using two different NCTM *Standards*-based curricula, equaled or exceeded the achievement of students from matched comparison districts, using different curriculum materials, on the state achievement test. The use of NCTM *Standards*–based materials and instruction grounded in learning preferences of all students can give more students access to high-quality mathematical knowledge.

CONCLUSION

Fundamental to progress in the mathematics learning of all students is the understanding by mathematics educators in grades K–16 schools that equity and access are achieved when all students have the opportunity for, and access to, mathematical understanding. Understanding comes when educators use the knowledge of students' learning in their practice. There are numerous ways to operationalize instructional accommodations for the development of students' mathematical understanding mentioned earlier in this article. Although I generally avoid itemized recommendations for instructional strategies, I am including recommendations here with the caveat that instructional strategies vary and are personally devel-

Learning Preferences	Pedagogical Accommodations	NCTM Outcomes in Mathematics Teaching and Learning
Students	Teachers should	Students should
Focus on the whole **Use improvisational, intuitive thinking**	• acknowledge and use individual students' preferences in the acquisition of knowledge.	• use mathematical reasoning, conjecturing, inventing, and problem solving; • have teachers who value differences to help everyone learn mathematics.
Learn from informal class discussion	• develop activities that promote mathematical discourse within the classroom among students and between the students and teacher.	• experience classrooms as mathematical communities; • learn to communicate and reason mathematically.
Recall relevant verbal ideas **Narrate human concepts**	• value students' discourse and verbal knowledge.	• employ logic and mathematical evidence as verification for answers; • connect mathematics, its ideas, and its applications.
Respond to extrinsic motivation	• encourage, support, and provide feedback to students as they learn; • expect that students can and will achieve conceptual and procedural understanding of the mathematics content.	• have resources and support; • become confident in their own ability; • experience teachers who have high expectations and provide worthwhile opportunities for all.
Focus on interests **Engage in human and social content material**	• create and use mathematical tasks that require students to "do mathematics."	• learn to value mathematics; • become mathematical problem solvers.
Achieve interdependently	• create interdependent learning communities within the classroom.	• experience classrooms as mathematical communities.

Figure 1.1. Comparison of learning preferences and recommended NCTM practices

oped by teachers through a process of reflection and profound thinking about the development and learning of their students and their personal epistemology.

On the basis of my call for a shift in thinking about how large numbers of students learn mathematics and a restructuring of instruction to address their learning preferences, I describe general recommendations for that restructured instruction.

1. Instruction should purposefully use students' improvisational thinking in classroom discussions. This means that teachers listen to students' thoughts and musings that do not seem to be related to the topic at first glance but after deeper questioning become apparent. Often this questioning reveals that students were on a trajectory to reach appropriate mathematical conclusions.
2. Instruction should have students narrate human events by creating problems based on life experiences that exemplify concepts they are learning or by solving problems that are related to struggles students and their families encounter economically and socially, such as the placement of toxic dump sites in low-income areas, the redlining of insurance rates, mortgages in low-income and minority areas, and so on.
3. Instruction should present a variety of strategies and approaches to solve problems and should welcome correct solutions that are original. Students should be allowed to solve problems holistically and analytically and support their conclusions through a spectrum of means including guess-and-check verifications and algorithms.
4. Instruction should encourage students to work on projects that require interdependency—the use of others' knowledge.
5. Instruction should assess students through methods that require verbal and written demonstration of knowledge.
6. Instruction should allow students to demonstrate knowledge through small and large productions that provide opportunities for the teacher, students, and parents to laud and acknowledge students' achievement.

In addition to these recommendations, instruction should offer students opportunities to learn, appropriate levels of challenge, and opportunities to take risks as they learn. We have to move from a narrow vision of teaching and learning mathematics where some students learn and understand and others are left behind, dissociated from mathematics learning. Mathematics instruction can and should offer all students the opportunity to be successful in learning and understanding mathematics. If excellence in achievement for all students is possible, and I believe that it is, then being good at educating some is not enough.

References

Briars, Diane J., and Lauren B. Resnick. "Standards, Assessments—and What Else? The Essential Elements of Standards-Based School Improvement." Report to the Pittsburgh School System, 2000.

Carpenter, Thomas P., and Richard Lehrer. "Teaching and Learning Mathematics with Understanding." In *Mathematics Classrooms That Promote Understanding,* edited by Elizabeth Fennema and Thomas A. Romberg, pp. 19–32. Mahwah, N.J.: Lawrence Erlbaum Associates, 1999.

Cobb, Paul, Terry Wood, Erna Yackel, and Betsy McNeal. "Characteristics of Classroom Mathematics Traditions: An Interactional Analysis." *American Educational Research Journal* 29 (fall 1992): 573–604.

Dance, Rosalie. "Modeling: Changing the Mathematics Experience in Post-Secondary Classrooms." Paper presented at The Nature and Role of Algebra in the K–14 Curriculum: A National Symposium, Washington, D.C., May 1997.

Gay, Geneva. *Culturally Responsive Teaching: Theory, Research, and Practice.* New York: Teachers College Press, 2000.

Grouws, Douglas A., and Linda O. Lembke. "Influential Factors in Student Motivation to Learn Mathematics: The Teacher and Classroom Culture." In *Motivation in Mathematics,* edited by Martha Carr, pp. 39–62. Cresskill, N.J.: Hampton Press, 1996.

Hale-Benson, Janice. *Black Children: Their Roots, Culture, and Learning Styles.* Baltimore: Johns Hopkins Press, 1986.

Harris, Olita D. "Equity in Classroom Assessment." In *Teaching from a Multicultural Perspective,* edited by Helen Roberts, Juan C. Gonzales, Olita D. Harris, Delores J. Huff, Ann M. Johns, Ray Lou, and Otis L. Scott, pp. 77–90. Thousand Oaks, Calif.: Sage, 1994.

Hilliard, Asa G. "Alternatives to IQ Testing: An Approach to the Identification of Gifted 'Minority' Children." Sacramento, Calif.: California State Department of Education, 1976. ERIC Document Reproduction Service No. ED147009.

Jonsson, Patrick. "Achievement Gap Narrows as Attitudes Change." *Christian Science Monitor,* December 18, 2001. www.csmonitor.com/2001/1218/p13s1-lecs.html.

Kozol, Jonathan. *Savage Inequalities.* New York: Crown, 1991.

Malloy, Carol E. "Including African American Students in the Mathematics Community." In *Multicultural and Gender Equity in the Mathematics Classroom: The Gift of Diversity,* 1997 Yearbook of the National Council of Teachers of Mathematics (NCTM), edited by Janet Trentacosta, pp. 23–33. Reston, Va.: NCTM, 1997.

Malloy, Carol E., and William W. Malloy. "Issues of Culture in Mathematics Teaching and Learning." *Urban Review* 30 (1998): 245–57.

Moses, Robert P., and Charles E. Cobb. *Radical Equations.* Boston: Beacon Press, 2001.

National Council of Teachers of Mathematics (NCTM). *Curriculum and Evaluation Standards for School Mathematics.* Reston, Va.: NCTM, 1989.

———. *Professional Standards for Teaching Mathematics.* Reston, Va.: NCTM, 1991.

———. *Principles and Standards for School Mathematics.* Reston, Va.: NCTM, 2000.

National Science Foundation. "Big City Students Make Gains in Math and Science." NSF PR 01-53. June 28, 2001.

Pilburn, Michael, Daiyo Sawada, Jeff Turley, Kathleen Falconer, Russell Benford, and Irene Bloom. *Reformed Teaching Observation Protocol, Reference Manual.* Tempe, Ariz.: Arizona State University, 2000.

Reys, Robert, Barbara Reys, Richard Lapan, Gregory Holliday, and Deanna Wasman. "Assessing the Impact of *Standards*-Based Middle Grades Mathematics Curriculum Materials on Student Achievement." *Journal for Research in Mathematics Education* 34 (January 2003): 74–95.

Schoenfeld, Alan. "Making Mathematics Work for All Children: Issues of Standards, Testing, and Equity." *Educational Research* 31 (January/February 2002):13–25.

Shade, Barbara. "The Influence of Perceptual Development on Cognitive Style: Cross Ethnic Comparisons." *Early Child Development and Care* 15 (1989): 137–55.

Shafer, Mary C. "Instructional Quality in the Context of Reform." Paper presented at the Research Presession of the Annual Meeting of the National Council of Teachers of Mathematics, Orlando, Fla., April 2001.

Sowder, Judith, and Randolph Phillipp. "Promoting Learning in Middle-Grades Mathematics." In *Mathematics Classrooms That Promote Understanding,* edited by Elizabeth Fennema and Thomas A. Romberg, pp. 89–108. Mahwah, N.J.: Lawrence Erlbaum Associates, 1999.

Steele, Claude. "Race and the Schooling of Black Americans." *Atlantic Monthly* 269 (April 1992): 67–78.

Stein, Mary Kay, Margaret Schwan Smith, Marjorie A. Henningsen, and Edward A. Silver. *Implementing Standards-Based Mathematics Instruction: A Casebook for Professional Development.* New York: Teachers College Press, 2000.

Stiff, Lee V. "African-American Students and the Promise of the *Curriculum and Evaluation Standards.*" In *Teaching and Learning Mathematics in the 1990s,* 1990 Yearbook of the National Council of Teachers of Mathematics (NCTM), edited by Thomas J. Cooney, pp. 152–58. Reston, Va.: NCTM, 1990.

U.S. Department of Education. *The Longitudinal Evaluation of School Change and Performance (LESCP) in Title I Schools Final Report.* Washington, D.C.: U.S. Department of Education, 2001.

Willis, Madge G. "Learning Styles of African American Children: A Review of the Literature and Interventions." In *African American Psychology: Theory, Research, Practice,* edited by A. Kathleen H. Burlew, W. Curtis Banks, Harriette P. McAdoo, and Daudi A. Azibo, pp. 260–78. Newbury Park, Calif.: Sage, 1992.

2

Developing a Well-Articulated Algebra Curriculum

Examples from the NCTM Academy for Professional Development

Jennifer M. Bay-Williams

Elizabeth M. Skipper

Susan K. Eddins

ALGEBRAIC thinking is one of the important content areas that students should be learning throughout school. It begins in kindergarten with recognizing simple patterns and continues through high school with analyzing functions. In looking across the grade bands for any given content strand—for example, algebra—it is apparent that experiences in prekindergarten–grade 2 are essential building blocks for grades 3–5 and so on through high school. This article shows how the algebra strand may develop across the grades. We begin with a middle school algebra exploration and the opportunities it offers to develop important mathematics. Then we share investigations from grades pre-K–2 and 3–5 that illustrate the earlier, related concepts to be developed in the elementary school years. Finally, we share a high school task that builds on the original focus task. These tasks are among the ones used in the Algebra Workshops for the National Council of Teachers of Mathematics (NCTM) Academy for Professional Development. Before introducing the tasks, we briefly describe the Workshops.

Professional development Workshops delivered by the NCTM Academy are designed to enable teachers, administrators, and other stakeholders to understand both the broad vision of teaching in grades K–12 and the more focused vision within a grade band. The goals of the Academy are to help participants (1) gain a clearer and deeper understanding of the meaning and implications of NCTM's *Principles and Standards for School Mathematics*

(NCTM 2000); (2) develop the knowledge and skills necessary to implement, and advocate for, high-quality mathematics instruction; (3) expand problem-solving skills, learn to overcome obstacles, and implement positive change in the classroom, school, and district; and (4) develop a personal "plan of action" to guide participants toward realizing the vision of the *Standards* when they return to their local schools. (For more information on the Academy, visit www.nctm.org/academy or send an e-mail request to academy@nctm.org.)

As Academy developers and facilitators, we acknowledge the crucial need for understanding specific grade level goals within the context of a well-articulated grades K–12 curriculum. Therefore, although participants enroll in grade-band-specific sessions (e.g., 6–8), attention is given to their grade band as well as related expectations at other levels.

ALGEBRAIC THINKING IN GRADES PRE-K–12

The NCTM *Principles and Standards for School Mathematics* addresses the understanding of patterns, relations, and functions in the first of four fundamental components for the Algebra strand. Specific grade-level expectations for this component are presented in figure 2.1.

Patterns, Relations, and Functions in Middle School

Academy presenters have used the Building with Toothpicks problem from *Navigating through Algebra in Grades 6–8* (Friel et al. 2001) with students and with Academy participants to illustrate the way in which patterns, relations, and functions are developed in the middle school. Here we describe the activity using the word *student* broadly to include middle school students and Academy participants. With a box of toothpicks at every table, students build the first four shapes in the growing pattern (see fig. 2.2).

Students are encouraged to study the construction and look for patterns in the number of toothpicks used for the perimeter. They are to use the patterns they notice to predict the number of toothpicks that will be needed for the fifth design in the pattern. Students build the fifth shape to assess the correctness of their prediction. Finally, students describe a general rule to find the perimeter for any design in this pattern. Before reading on, look at figure 2.2 and solve the problem using a strategy of your choice.

Students approach the problem in different ways. With each approach, they are analyzing, representing, and generalizing the pattern. The sharing of solution strategies for this task is very exciting because there are so many ways to solve it. One approach is to record the information in a table. With each new shape, four more toothpicks are needed (a recursive pattern). Comparing the shape number to the number of toothpicks, students note

Grades Pre-K–2

- Sort, classify, and order objects by size, number, and other properties
- Recognize, describe, and extend patterns such as sequences of sounds and shapes or simple numeric patterns and translate from one representation to another
- Analyze how both repeating and growing patterns are generated

Grades 3–5

- Describe, extend, and make generalizations about geometric and numeric patterns
- Represent and analyze patterns and functions, using words, tables, and graphs

Grades 6–8

- Represent, analyze, and generalize a variety of patterns with tables, graphs, words, and, when possible, symbolic rules
- Relate and compare different forms of representation for a relationship
- Identify functions as linear or nonlinear and contrast their properties from tables, graphs, or equations

Grades 9–12

- Generalize patterns using explicitly defined and recursively defined functions
- Understand relations and functions and select, convert flexibly among, and use various representations for them
- Analyze functions of one variable by investigating rates of change, intercepts, zeros, asymptotes, and local and global behavior
- Understand and perform transformations such as arithmetically combining, composing, and inverting commonly used functions, using technology to perform such operations on more-complicated symbolic expressions
- Understand and compare the properties of classes of functions, including exponential, polynomial, rational, logarithmic, and periodic functions
- Interpret representations of functions of two variables

Fig. 2.1. Expectations in grades pre-K–12 related to patterns, relations, and functions

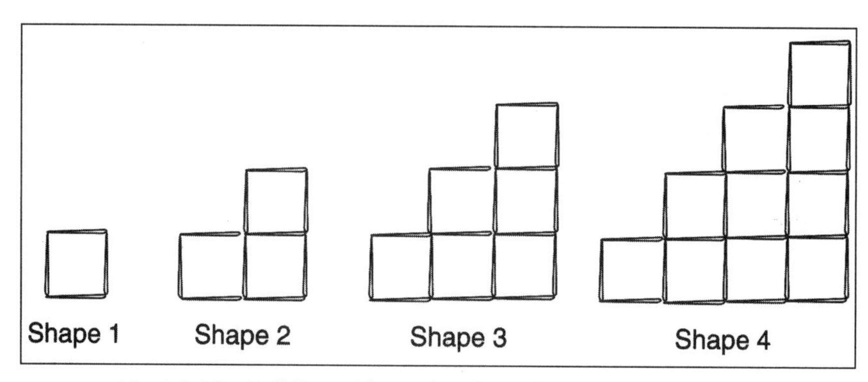

Fig. 2.2. The Building with Toothpicks problem (Friel et al. 2001)

that the number of toothpicks needed is four times whatever the shape number is (an explicit pattern). Symbolically, students note that the pattern is "4 times n," or "$4n$" or "$4(n)$" (see fig. 2.3). Tabular data can also be entered into a graphing calculator or graphing software. Students can graph the function and see on the graph a relationship between design number and number of toothpicks needed for the perimeter.

There are many ways that students have used the toothpick construction without a table to generalize the pattern. One way is to note that on design 3 there are 3 toothpicks for the base of the design and 3 for the right side. Also, there are 3 "steps" and each "step" uses 2 toothpicks, so 6 toothpicks are needed for the steps. Combining the three parts of the design (base, right side, and steps) results in 12 total toothpicks. In symbols, this pattern is $n + n + 2n$. Another way to look at the shapes is to consider that for shape 3, there are 3 horizontal toothpicks on the bottom and 3 on the top. There are

Shape Number	Number of Toothpicks in Perimeter
1	4
2	8
3	12
4	16
5	20
10	40
	4 times n
n	$4n$
	$4(n)$

Fig. 2.3. Table of data for toothpick problem

also 3 vertical toothpicks forming the steps and 3 forming the right side. Therefore there are 3×2 vertical toothpicks and 3×2 horizontal toothpicks, so for any sized shape, $2(n) + 2(n)$ toothpicks form the perimeter. A third strategy involves changing the shape of the "staircases." If the "steps" are dismantled, the toothpicks used to create the perimeter can be reused to form a square. That leads to the general case that it will take $4(n)$ toothpicks for any sized shape. As students share their observations, others see the pattern they noticed and can make the connections between the symbols and the picture. Teachers can also engage students in a rich discussion about the different symbolic expressions and ask students to explain or show that they are equivalent.

Building a geometric growing design offers students the opportunity to analyze the physical model, the table where they record their data, and the graph they might create. Middle school students are expected to describe the general situation in words and symbols. Students should also recognize that there are multiple ways to describe a pattern, and they should be able to determine when one expression is equivalent to another. Developing students' understanding of multiple symbolic representations is supported by the geometric models.

The Building with Toothpicks problem is an excellent way to begin the study of patterns, relations, and functions in middle school because the general instance, or function, is a multiple of the shape number ($y = mx$) and is therefore a good entry into using and comparing symbolic rules. Extensions to this problem include other toothpick patterns, finding the total number of toothpicks, and finding the number of squares formed. This model of investigation can be applied to more difficult problems to generalize. For example, another activity found in the grades 6–8 Algebra Navigations book is Stacking Cups, which involves finding the formula for the height of a stack of n cups. The formula for this pattern is in the form $y = mx + b$, making it a more challenging pattern for students to analyze and find the general rule.

Patterns in Grades Pre-K–2

Although looking at tables, graphs, and symbolic rules for growing patterns is a middle school expectation, students' experiences with growing patterns begin in their very first years of school. Primary school students explore all kinds of patterns, including colors, shapes, and numbers. Patterns may be repeating or growing. In some patterns, such as 2, 4, 6, __, the next term could be 8 if it was a plus 2 pattern, it could be 2 if it was a repeating pattern with 2, 4, 6 as the core, or it could be any other number. Primary school students can predict what comes next and offer their explanation for why they think so. (See Bay-Williams 2001 for elaboration on this topic.) Students also begin to look at pictures and models and how they grow. In the case of a geometric growing pattern, there is often just one possibility for the next shape in the pattern. The following task, from How Does It Grow? in *Navigating through Algebra in Prekindergarten–Grade 2* (Greenes et al. 2001), is shared in order to consider the continuum of experiences through the grades. Figure 2.4 shows two different patterns.

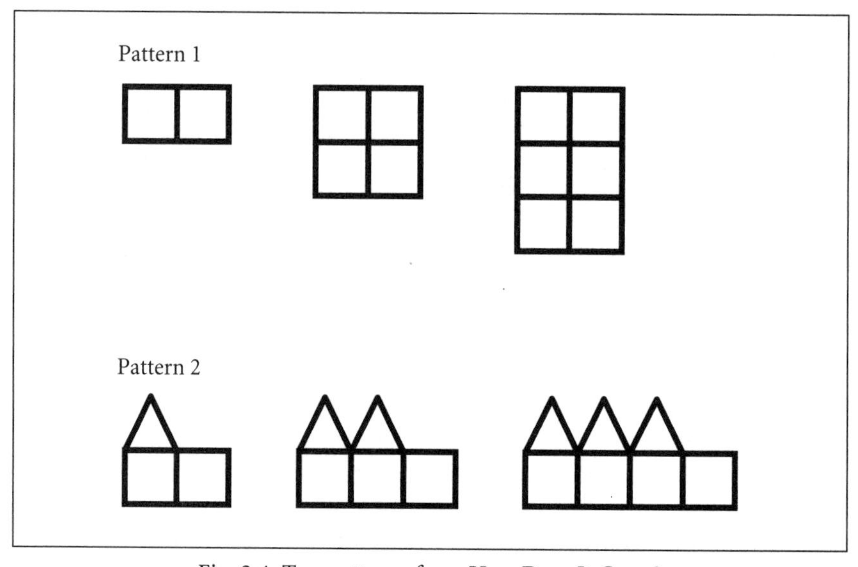

Fig. 2.4. Two patterns from How Does It Grow?

Students explore pattern 1, by describing what is alike and different from one design to the next. As the teacher points at the first, second, and third design, she asks, "How is this pattern growing?" Students explain that there are two more squares, or another row of squares is added on the bottom. Students predict how many squares will be in the next pattern. Students can be asked if it is possible to have 7 squares in one of the designs. The second pattern, which incorporates two different shapes, presents an opportunity to ask more challenging questions. The teacher could begin by asking students what they think the next figure will look like. Students can also predict how many squares will be in a design that has 6 triangles. Conversely, students can find out how many triangles are needed in a shape that has 10 squares.

The visible geometric pattern enables students to see how a pattern can grow and to study the way in which it is growing. Primary school students describe in words that pattern 1 "goes up by two squares," or pattern 2 "adds one square and one triangle each time." Students demonstrate their understanding by building the shapes with pattern blocks and drawing the figures. Students use the constructions to predict how many will be needed for the next design in the sequence. Skills developed here (analyzing patterns, predicting what comes next, and describing the pattern in words) provide a foundation for the thinking needed for the toothpick problem. The ability to describe how to find the next term is an important building block in eventually being able to find the generalization and to describe the general case symbolically.

Studying Patterns in Grades 3–5

Grades 3–5 students need many opportunities to explore patterns in greater depth in order to make conjectures about patterns and then test those conjectures. "Tiling a Patio," found in *Navigating through Algebra in Grades 3–5* (Cuevas and Yeatts 2001, pp. 18–22), is an excellent investigation to illustrate this point. The following scenario is used to engage the students:

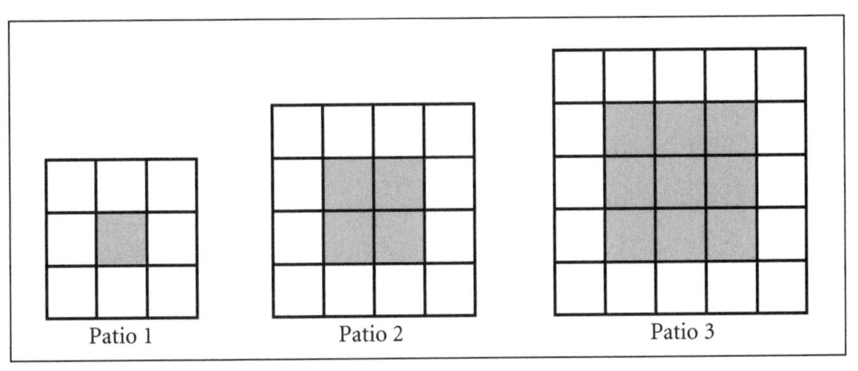

Fig. 2.5. Patio pattern

> Alfredo Gomez is designing square patios. Each patio has a square garden area in the center. Alfredo uses brown tiles to represent the soil of the garden. Around each garden, he designs a border of white tiles. The pictures [see fig. 2.5] show the three smallest square patios that he can design with brown tiles for the garden and white tiles for the border.

Students are asked to construct the three patios and record the number of brown and white tiles for each patio using a table. Students study the designs and the related tables to analyze the growing pattern. Using these multiple representations, students can verbalize recursive patterns they notice in each column, such as "the white tiles go up by four each time" and they can describe in words some generative patterns for each column (see fig. 2.6).

With the Tiling a Patio investigation, students make connections between the physical representation of the patterns they have constructed and their verbal or graphical descriptions. In grades 3–5, students need experiences with describing, extending, and making generalizations about geometric and

Patio Number	Number of Brown Tiles	Number of White Tiles	Total Number of Brown and White Tiles
1	1	8	9
2	4	12	16
3	9	16	25
patio number	“Patio number times itself” “Patio number times the patio number”	“Four times the patio number plus 4” “Patio number plus two and double that. Then add two times the patio number.” “Find the area of the outside square and subtract the area of the inside square.”	“Add 2 to the patio number and multiply it by itself.”

Fig. 2.6. Table for the patio problem

numeric patterns. The geometric patterns at this level are more sophisticated than those in the primary grades, and at this level, students are introduced to multicolumned tables and graphs to illustrate and analyze the growing patterns. Elementary school students are able to describe the growth pattern in words, though they may not be able to describe it in variables, as they will in the middle grades.

High School Patterns and Functions

Just as experiences in the intermediate grades support the development of more sophisticated algebraic thinking in middle school, the middle school experiences prepare students for high school topics. Recall the Building with Toothpicks problem. Students were able to represent the problem physically, in a table, graph, words, and symbols. Being able to understand and use flexibly multiple representations prepares students to solve much more challenging growth patterns in high school.

High school teachers may not be used to working with manipulatives, thinking that students should move to greater levels of abstraction by the time they have completed eighth grade. Much is to be gained, however, by encouraging students to build on the tactile experiences they have had in middle school and earlier as they build bridges to algebraic concepts. The grades 9–12 Algebra Workshops use a three-dimensional pattern called the Skeleton Tower posted on PBS Mathline (2003).

Although this problem may be introduced at the middle school, in high school it provides an excellent introduction to summation processes, symbolic representation, and quadratic functions.

In the Skeleton Tower pattern, the first shape is one cube (the top of the tower); the second shape uses a total of six cubes, with a new, larger layer added on beneath the previous layer. Figure 2.7 is an illustration of the sixth tower.

Students construct towers with various center heights with cubes. As they build, they are encouraged to describe the tower in as many ways as they can. Like the toothpick problem, there are multiple ways to decompose the geometric model, each of which leads to a variation in the initial symbolic representation of the problem. One approach is to view the tower as four sets of steps with one center tower. That leads to the following symbolic notation:

$$4(1 + 2 + 3 + \ldots (n - 1)) + n$$

Another way students view the problem is to see two staircases, each one going up and then down. The two staircases are set at right angles to each other, and the center stack is counted in both and therefore must be subtracted. The resulting formula is

$$2[(1 + 2 + 3 + \ldots + (n - 1) + n + (n - 1) + \ldots + 3 + 2 + 1)] - n.$$

Fig. 2.7. The sixth skeleton tower

Some students describe the number of blocks in each horizontal layer of the tower. The top layer has 1 block, the next layer 5 blocks, the next 9 blocks, and so on. Symbolically, that pattern is $(1 + 5 + 9 + \ldots + (4n + 1))$. Another technique is to take one "staircase" or "arm" and turn it over to "fit" on top of another "arm." That transforms the staircase into a central tower and two rectangles, each with base n and height $(n - 1)$, leading to the representation of $2(n(n - 1)) + n$. A slight modification of this approach is to form two squares, n blocks on a side, each of which includes the center tower and leads to the formula $2(n)(n) - n$. As high school students work to reconcile these multiple representations of what must ultimately be the same number of blocks, algebraic formulas (e.g., sum of the first n integers), simple symbolic manipulation involving exponents, and the distributive property actually become useful tools instead of sterile rules.

High school students, like younger students, should be encouraged to create tables that show the number of blocks needed to build towers of different heights (see fig. 2.8). These tables can be used to explore first- and second-order differences.

The fact that second-order differences are constant indicates that the number of blocks follows a quadratic pattern of growth. Specifically, the integral pairs that are generated by this function all lie on the graph of a parabola with vertex (1/4,–1/8) and x-intercepts 0 and 1/2. Other examples can follow where the pattern of growth is exponential, allowing for comparisons of these two basic forms of nonlinear growth. By grounding explorations in experiences with materials or stories, students are afforded concrete as well

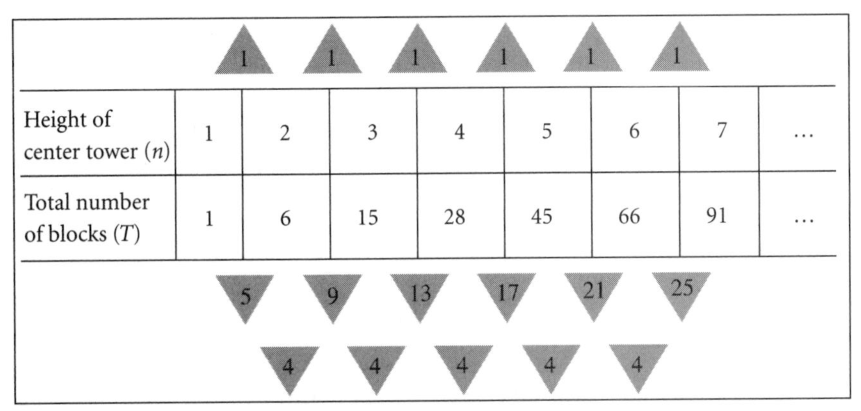

Height of center tower (n)	1	2	3	4	5	6	7	...
Total number of blocks (T)	1	6	15	28	45	66	91	...

Fig. 2.8. Table of first and second differences for the Skeleton Tower pattern

as abstract ways of thinking about the situation and solving the problem, enabling and encouraging more students to be successful in understanding algebra.

SUMMARY

The experiences described above illustrate how ideas must build throughout a student's grades pre-K–12 education in a well-articulated curriculum. Imagine how students would perform on the Skeleton Tower problem if they had not had experiences such as How Does It Grow?, Tiling a Patio, and Building with Toothpicks. The expectations for students change and become more advanced with each level of task. A well-articulated curriculum must also focus on important mathematics, and it must be coherent. The tasks described here focus on looking at relationships among quantities, representing mathematical relationships in different ways, and analyzing change, unlike the symbolic manipulation that has historically dominated the algebra curriculum. Students can make connections within algebra and to other strands as they use geometric models to analyze patterns. A curriculum that integrates algebra and geometry, as well as other mathematics strands, presents an interconnected vision of mathematics. Exploring patterns in different contexts is also a way for students to see mathematics in their world. It is important to note that the explorations described above are not intended to be single tasks taught in isolation but should be embedded in meaningful ways and build on students' prior experiences. "A curriculum is more than a collection of activities: it must be coherent, focused on important mathematics, and well articulated across the grades" (NCTM 2000, p. 14).

An important tenet of the Teaching Principle is that teachers have "deep, flexible knowledge about curriculum goals and about the important ideas that are central to their grade level" (NCTM 2000, p. 17). Teachers must understand the mathematics that is taught previous to the level they teach, so that they can effectively build on those experiences. A lack of awareness can result in too much repetition or gaps in students' learning. Similarly, teachers need to know what is coming in subsequent years so they can build foundational ideas to support students' thinking at the next level. Professional development, such as the NCTM Academy, must not stop at preparing teachers for their grade level alone but must prepare them with a broader perspective of where their students have been and where they will be going. A well-articulated curriculum and worthwhile tasks set the stage for effective teaching, but alone they are not enough:

> Teachers must also decide what aspects of the task to highlight, how to organize and orchestrate the work of the students, what questions to ask to challenge those with varied levels of expertise, and how to support students without taking over the process of thinking for them. (NCTM 2000, p. 19)

Having a well-articulated curriculum focusing on interconnectedness and applications increases the chances that teachers can implement the Teaching Principle and that students will have a deeper, more comprehensive understanding of mathematics.

REFERENCES

Bay-Williams, Jennifer M. "What Is Algebra in Elementary School?" *Teaching Children Mathematics* 8 (December 2001): 196–200.

Cuevas, Gilbert J., and Karol Yeatts. *Navigating through Algebra in Grades 3–5.* Reston, Va.: National Council of Teachers of Mathematics, 2001.

Friel, Susan, Sid Rachlin, and Dot Doyle, with Claire Nygard, David Pugalee, and Mark Ellis. *Navigating through Algebra in Grades 6–8.* Reston, Va.: National Council of Teachers of Mathematics, 2001.

Greenes, Carol, Mary Cavanagh, Linda Dacey, Carol Findell, and Marian Small. *Navigating through Algebra in Prekindergarten–Grade 2.* Reston, Va.: National Council of Teachers of Mathematics, 2001.

National Council of Teachers of Mathematics (NCTM). *Principles and Standards for School Mathematics.* Reston, Va.: NCTM, 2000.

PBS Mathline. "Algebraic Thinking Math Project: Looking through the Algebraic Lens, Grades 6–8." www.pbs.org/teachersource/mathline/lessonplans/pdf/atmp/AlgebraicLens68.pdf, June 9, 2003.

3

What Makes a Mathematical Task Worthwhile?

Designing a Learning Tool for High School Mathematics Teachers

Fran Arbaugh

Catherine A. Brown

> In effective teaching, worthwhile mathematical tasks are used to introduce important mathematical ideas and to engage and challenge students intellectually. Well-chosen tasks can pique students' curiosity and draw them into mathematics. ... Regardless of the context, worthwhile tasks should be intriguing, with a level of challenge that invites speculation and hard work.
>
> —National Council of Teachers of Mathematics, *Principles and Standards for School Mathematics*

For decades educators have been exploring the relationship of tasks, intellectual challenge, and students' learning. For example, Doyle (1983) argued that "tasks form the basic treatment unit in classrooms" (p. 162) and that tasks "are defined by the answers students are required to produce and the routes that can be used to obtain these answers" (p. 161). Further, given that different tasks require different intellectual demands, he argued that if all the tasks used in a classroom require a particular type of answer and route, then students' learning will be limited to that type of answer obtained by taking that particular route.

Doyle's work on academic tasks was not bound by a content discipline: he wrote about tasks as they occur across the curriculum. A "translation" of Doyle's work with regard to teaching and learning mathematics is provided by Stein and Smith (1998, p. 269):

> Tasks that ask students to perform a memorized procedure in a routine manner lead to one type of opportunity for student thinking; tasks that require students

> to think conceptually and that stimulate students to make connections lead to a different set of opportunities for student thinking.

Mathematical tasks form the core of activity in the high school mathematics classroom. It is important for teachers in those classrooms to consider the types of opportunities to learn mathematics that they provide for their students:

> Students learn through the experiences that teachers provide. Thus, students' understanding of mathematics, their ability to use it to solve problems, their confidence in, and disposition toward, mathematics are all shaped by the teaching they encounter in school. (National Council of Teachers of Mathematics [NCTM] 2000, pp. 16–17)

Once mathematics textbooks are adopted at the district or school level, teachers often have sole responsibility for choosing the mathematical tasks they have their students complete. Many high school teachers use only problems that are found in their textbooks, whereas other teachers may use a combination of problems from their textbooks and outside sources. Because the teachers are often individually choosing tasks for their students to complete, it becomes important that high school teachers understand characteristics of "worthwhile mathematical tasks" (NCTM 1991). In turn, it becomes important that the mathematics education community support high school teachers in ways that encourage them to think deeply about the tasks that they have their students complete.

Our goal in this article is to share information about a learning tool that we designed to support high school mathematics teachers in thinking deeply about what constitutes worthwhile mathematical tasks. The learning tool we designed is a task-sorting activity, and we based our design in work done by the QUASAR (Quantitative Understanding: Amplifying Student Achievement and Reasoning) project (Stein et al. 2000). In their work with middle schools in economically disadvantaged communities, the QUASAR project created a set of criteria for categorizing mathematical tasks on the basis of the type(s) of thinking that the task requires of students. They call this set of criteria the "Levels of Cognitive Demand." They then designed an activity to use with teachers to help them learn about the Levels of Cognitive Demand.

Figure 3.1 contains the Levels of Cognitive Demand criteria. These criteria were developed by the QUASAR researchers after a close analysis of hundreds of tasks used in project classrooms—criteria that "when applied to a mathematical task (in print form) ... can serve as a judgment template (a kind of scoring rubric) that permits a 'rating' of the task based on the kind of thinking it demands of students" (Stein et al. 2000, p. 15).

Levels of Demands

Lower-level demands (memorization)

- Involve either reproducing previously learned facts, rules, formulas, or definitions or committing facts, rules, formulas or definitions to memory.
- Cannot be solved using procedures because a procedure does not exist or because the time frame in which the task is being completed is too short to use a procedure.
- Are not ambiguous. Such tasks involve the exact reproduction of previously seen material, and what is to be reproduced is clearly and directly stated.
- Have no connection to the concepts or meaning that underlie the facts, rules, formulas, or definitions being learned or reproduced.

Lower-level demands (procedures without connections)

- Are algorithmic. Use of the procedure either is specifically called for or is evident from prior instruction, experience, or placement of the task.
- Require limited cognitive demand for successful completion. Little ambiguity exists about what needs to be done and how to do it.
- Have no connection to the concepts or meaning that underlie the procedure being used.
- Are focused on producing correct answers instead of on developing mathematical understanding.
- Require no explanations or explanations that focus solely on describing the procedure that was used.

Higher-level demands (procedures with connections)

- Focus students' attention on the use of procedures for the purpose of developing deeper levels of understanding of mathematical concepts and ideas.
- Suggest explicitly or implicitly pathways to follow that are broad general procedures that have close connections to underlying conceptual ideas as opposed to narrow algorithms that are opaque with respect to underlying concepts.
- Usually are represented in multiple ways, such as visual diagrams, manipulatives, symbols, and problem situations. Making connections among multiple representations helps develop meaning.

- Require some degree of cognitive effort. Although general procedures may be followed, they cannot be followed mindlessly. Students need to engage with conceptual ideas that underlie the procedures to complete the task successfully and that develop understanding.

Higher-level demands (doing mathematics)

- Require complex and nonalgorithmic thinking — a predictable, well-rehearsed approach or pathway is not explicitly suggested by the task, task instructions, or a worked-out example.
- Require students to explore and understand the nature of mathematical concepts, processes, or relationships.
- Demand self-monitoring or self-regulation of one's own cognitive processes.
- Require students to access relevant knowledge and experiences and make appropriate use of them in working through the task.
- Require students to analyze the task and actively examine task constraints that may limit possible solution strategies and solutions.
- Require considerable cognitive effort and may involve some level of anxiety for the student because of the unpredictable nature of the solution process required.

Source: Smith and Stein 1998, p. 348

Fig. 3.1. Levels of Cognitive Demand

HELPING TEACHERS LEARN ABOUT THE LEVELS OF COGNITIVE DEMAND

The work that the QUASAR project did with regard to mathematical tasks and cognitive demand piqued our interest. As mathematics educators working with preservice and in-service mathematics teachers, we had struggled with a way to help mathematics teachers conceptualize and verbalize the characteristics of worthwhile mathematical tasks (NCTM 1991). We were particularly intrigued with a task-sorting activity that the QUASAR project developed to help middle school teachers learn about the Levels of Cognitive Demand criteria. As Stein et al. (2000, p. 18) explain:

> One way we have found to help teachers learn to differentiate levels of cognitive demand is through the use of a task-sorting activity. The long-term goal of this activity is to raise teachers' awareness of how mathematical tasks differ with respect to their levels of cognitive demand, thereby allowing them to better match tasks to goals for student learning.

After learning about the task-sorting activity, we obtained it and used the activity with a group of in-service middle school teachers during a professional development session. We found that the teachers categorized many of the tasks differently, and those differences supported interesting and insightful whole-group discussion. On the basis of our experience with these middle school mathematics teachers, we wanted to engage our high school preservice teachers in a similar experience. But if we wanted to use a task-sorting activity with high school teachers, we would need to create a task-sorting activity based in high school mathematics content.

Creating a High School–Based, Task-Sorting Activity

As we began gathering tasks for our card-sorting activity, we knew that we wanted to find tasks that supported the level of discussion we had experienced with the middle school teachers. We looked for high school tasks that would prompt similar discussion. For example, task N of the QUASAR set (categorized as "procedures without connections" by QUASAR, thus requiring a low level of cognitive demand; see fig. 3.2) tended to promote conversation about whether a task that is presented as a word problem is always considered to have a high level of cognitive demand. Task N also prompted conversation about the quality of explanation—whether explaining a procedure required a high level of cognitive demand, or if something more was required. We purposefully searched examples of tasks at the high school level that would have similar features. (See, for example, tasks F and G in the appendix to this article.) Although we had a sense of the types of tasks that we needed to include, we believed that we needed more guidance, and so we turned to the QUASAR project for that guidance.

Stein and her colleagues made a distinction between superficial and substantive features of tasks. Superficial features are those that appear to meet reform recommendations but do not do so in substantive ways. Examples of these types of superficial task features include that the task requires the use of a calculator or diagram or involves multiple steps to complete, or that the task requires a written explanation or has a real-world context.

Using those task features and others, Stein et al. (2000) presented results from a careful analysis of eight of the tasks they used in their task-sorting

Task N

The cost of a sweater at J. C. Penney's was $45.00. At the "Day and Night Sale" it was marked 30% off of the original price. What was the price of the sweater during the sale? Explain the process you used to find the sale price.

Fig. 3.2. Task N (Middle School)

activity. They analyzed those eight tasks (two from each level of cognitive demand) across several task features. They used this analysis to ensure themselves that certain task features, which often represent surface characteristics of tasks, varied across levels. For example, they wanted both low- and high-level tasks that require an explanation so that teachers would consider the cognitive demand required by different types of explanations.

We thought that a similar analysis would assure us that the tasks we had chosen varied in the manner described above. Once we had agreed on twenty tasks to include and were satisfied with the distribution of those tasks across levels of cognitive demand, we conducted an analysis of the twenty tasks on the basis of their features. We chose eight of those tasks and describe our analysis below (see the appendix for the tasks).

Table 3.1 presents the results of the analysis we conducted on eight of the tasks in the high school set. (The analysis we conducted on all twenty tasks can be found in Arbaugh [2000].) The overall structure of our analysis of the high school set of mathematical tasks is the same structure that Stein et al. (2000) used in analyzing the middle school set of tasks. The features included in the far right column are those used by Stein et al. (2000), and the explanation of categorization is grounded in the Levels of Demand criteria.

Note that many of the features describe tasks that are placed in different categories. For example, both task G and task J have real-world contexts but are categorized, respectively, as requiring lower-level cognitive demand (procedures without connections) and requiring higher-level cognitive demand (doing mathematics). Similarly, task F and task J both require explanations but are categorized at different levels of cognitive demand. Both task E and task K are textbooklike and require different levels of cognitive demand, with task E requiring a higher level of cognitive demand (procedures with connections) and task K requiring a lower level of cognitive demand (memorization).

We concluded that we had a set of tasks in which both the level of cognitive demand and the task characteristics varied across the other dimension.

IMPLEMENTING THE TASK-SORTING ACTIVITY WITH PRESERVICE AND IN-SERVICE HIGH SCHOOL MATHEMATICS TEACHERS

Implementing with Preservice Mathematics Teachers

We have used the high school–based task-sorting activity with six different groups of preservice high school mathematics teachers. On the basis of observation, assessments, and classroom-based artifacts, we believe that

Table 3.1
Cognitive Demand and Features of Eight Sample Tasks (High School)

Task	Level of Cognitive Demand	Explanation of Categorization	Features
A	Doing mathematics	No pathway is suggested by the task. The task requires students to access relevant knowledge and experiences. The task may involve some level of anxiety for the student.	• Is nonalgorithmic • May use a calculator
C	Memorization	The task involves reproducing previously learned definitions. The task has no connection to concepts or meaning.	• Is textbooklike
E	Procedures with connections	The task focuses students' attention on the use of procedures to develop deeper levels of understanding. Procedures cannot be followed mindlessly.	• Has real-world context • Is textbooklike • Involves multiple steps • May use a graphing calculator
F	Procedures without connections	Little ambiguity exists for what needs to be done. The task requires no connection to meaning. The task focuses the student on getting the right answer. The task requires an explanation that focuses solely on describing the procedure used.	• Is textbooklike • Requires an explanation • Uses a diagram • May use a calculator
G	Procedures without connections	The use of a procedure is specifically called for. The task has no connection to underlying mathematics. The focus is on obtaining the correct answer.	• Is textbooklike • Has real-world context • Involves multiple steps
J	Doing mathematics	No pathway is suggested by the task. The task requires students to access relevant knowledge and experiences. The task requires nonalgorithmic thinking.	• Involves multiple steps • Requires an explanation • Has real-world context
K	Memorization	The task involves reproducing previously learned rules and exact reproduction of previously learned terms.	• Is textbooklike • Is symbolic or abstract
L	Procedures with connections	The task focuses on using procedures to build deeper understanding. The task requires the use of multiple representations. Procedures cannot be followed mindlessly.	• Uses multiple representations • Involves multiple steps • May use a calculator

using the task-sorting activity to help the preservice teachers learn the Levels of Cognitive Demand criteria has been a successful endeavor. Each group has come away from their methods class with the ability to categorize tasks according to the criteria, and overall the preservice teachers have been able to communicate the importance of considering the cognitive level demanded of the tasks they will use in their future mathematics classrooms. Whether they carry that conviction into their classes is a subject for future study. We continue to use the task-sorting activity with preservice teachers, believing that it gives them both a vocabulary with which to discuss tasks and a way in which to analyze texts critically.

One of the ways we require our students to demonstrate their knowledge of the levels of cognitive demand is through a textbook analysis undertaken as a course assignment. We find that providing a language and a set of criteria with which to analyze the tasks contained in a textbook supports our students in analyzing textbooks in a critical manner. When we insist that our students focus on the cognitive demand of the textbook tasks, we have found that they focus less on *Standards*-based buzzwords, the physical layout of the text (whether it "looks like a mathematics text"), and the number of problems in a particular section. They have also learned to look past surface characteristics of the individual problems and analyze them on the basis of the types of thinking required by students. In the end, learning about the Levels of Cognitive Demand has helped them form a clearer vision of worthwhile mathematical tasks to use in their future classrooms.

Implementing with In-Service Mathematics Teachers

In the fall of 1999, we used this task-sorting activity with a group of high school geometry teachers who were meeting on a regular basis in a study group designed to support the teachers in enhancing their geometry students' reasoning abilities. The teachers were concerned that their geometry students were not developing good reasoning skills in their geometry classes—that the students left geometry class with a list of memorized vocabulary and rote proof-writing skills. We believed that engaging the teachers in a critical examination of the mathematical tasks they were using in their geometry classes would be a place to start them thinking about how to promote better reasoning in their students. The teachers agreed that this would be a good place to start.

One of the first activities in which the group participated was the task-sorting activity we had designed. They spent a majority of their first study group meeting sorting the tasks into four categories on the basis of the Levels of Cognitive Demand criteria. We began the follow-up discussion at that meeting and continued it into the second study group meeting. Although the task-sorting activity in itself is an important part of the learning experience, the follow-up discussion is equally important:

> The benefits of a task-sorting activity ... accrue not simply from completing the sort, but rather from a combination of small- and large-group discussions that provide the opportunity for discussion that moves back and forth between specific tasks and the characteristics of each category.... We have found that participants do no always agree with each other—or with us—on how tasks should be categorized, but that both agreement and disagreement can be productive. (Stein et al. 2000, p. 20)

Our experience with these high school teachers was similar. We focus here on two episodes from the large-group discussions: (1) the conversations prompted by task A and (2) the conversations that occurred as the teachers sought to distinguish between procedures with connections and procedures without connections.

Task A (see the appendix) prompted much discussion that we'll label "can't judge a book by its cover." The teachers initially categorized this task in three different levels of cognitive demand: "procedures without connections," "procedures with connections," and "doing mathematics." On the surface, task A looks rather benign, or as Ed described it: "It's short and all the words seem familiar. And it's a small number." "But," he added, "not necessarily an easy problem." Ed knew that task A was "not necessarily an easy problem" because he and his partner spent a good amount of time trying to solve it before categorizing it. On the basis of their experience, Ed and his partner found that they were collecting data, looking for patterns, and making and testing conjectures. Consequently, Ed and his partner categorized this problem as requiring a high level of cognitive demand and were able, during the large-group discussion, to convince their colleagues that their categorization was correct.

The conversation among the teachers in the study group about the cognitive demand required by task A led to more generalized conversation about the difficulty of choosing worthwhile tasks to use in class, a job made more difficult by the surface characteristics of the problems. It is interesting to note that of the multiple times we have implemented this activity, this discussion about the difficulty of choosing tasks for use in class was unique to the in-service high school teachers: it was a subject never brought up by the preservice teachers. The in-service teachers were immediately able to use their new knowledge about mathematical tasks to reflect on their own practice. The preservice teachers had no experience on which to reflect for choosing tasks to use in their classrooms.

Another topic that was prompted by our discussion of task A was the difference between a task that requires a high level of cognitive demand and a task that is difficult for students. Annie, a first-year teacher, appeared most troubled about wanting to understand the difference. She made the following statement during our discussion of task A:

> An interesting thing to note is that I don't necessarily think that difficulty or amount of time to do [a task] is necessarily a criterion of whether it's a high thinking level or not.... "What parts of your brain does it engage?" is a whole different thing. Are there multiple approaches? Do they really have to communicate and get the reasoning out, or can they just sit there and punch buttons on the calculator, and write an answer?

Annie kept coming back to the issue throughout our discussions, and although this issue was never resolved to the satisfaction of the group, it nonetheless prompted rich conversation that resulted in the teachers thinking deeply about the issue of a difficult task versus one that requires a high level of cognitive demand. The teachers were struggling with the ideas embedded in understanding worthwhile mathematical tasks. Is a task worthwhile just because the students find it difficult, or is there something more involved?

We also had an extended conversation about the difference between procedures without connections and procedures with connections. This discussion was prompted by Brian, who observed that the group seemed to have difficulty agreeing about the placement of several tasks with regard to these two categories. An exchange between Craig and Megan began to help clarify the confusion:

> *Craig:* I think it's the last part *[looking at the Levels of Cognitive Demands criteria]*. With the low level it says "requires no explanations" or just how you did it and not why.
>
> *Megan:* Task F was kind of that. You had to write a paragraph and explain how you found x.
>
> *Craig:* The question is "why?"
>
> *Megan:* "Why." Right.

Craig was arguing that even though this task asks the student to write a paragraph as part of the response, it is not at a high level of cognitive demand. He challenged Megan to see that writing about what a student does to solve a problem is different from writing about why that student chooses to do the what to solve the problem, with the latter requiring a higher level of thinking.

Although they agreed that Craig and Megan had found a difference, the teachers weren't convinced that it was that simple. Consider the following piece of transcript in which the teachers and the first author of this article were continuing the discussion started by Craig and Megan:

> *Fran:* I think it's hard because I can go through here and find bullets that I think would fit this [task F]. For example, the last one on procedures with connections. "Requires some degree of cognitive effort." I mean I think that this requires some degree of cognitive effort. I have to know something about the relationship of the angles, etc, etc.

Ed: What would be an example of a problem that didn't require cognitive effort?

Fran: What would be an example of a problem that wouldn't?

Wes: 2 + 3. Someone has learned to parrot that, but, of course, just as an example that doesn't require cognitive effort. It's not a problem unless it does.

Fran: So is it a degree of cognitive effort? Do you feel like there's a certain kind of cutoff about ...

Annie: No, different groups of students are going to have different kinds of cognitive effort. I mean normally you wouldn't think of vocabulary type of problems ... [like] giving them different types of diagrams with lines and transversals and identifying the various angles. Identify the transversal and the angle relationships in different diagrams. That seems pretty low level, because it's just basically based on memorized facts. But for people who are just trying to get a handle on these words, that's like trying to read a map in a foreign language.

Wes's lone comment in the piece of transcript above brought the concept of "problem" and "problematic" into the conversation. He was saying that a mathematics task does not qualify as a problem for students unless it requires some degree of cognitive effort—unless it is problematic for them. Annie picked up on that and tied it to the idea that cognitive effort involves considering what prior knowledge a student brings to a task. Shortly after the exchange above, Carl said, "I don't think there's anything wrong with having [the tasks] in different categories. It depends on what you teach and where you are." Carl was agreeing with Annie that the level of cognitive demand required by a mathematical task often is contingent on students' prior knowledge, and that teachers need to consider students' prior knowledge when choosing worthwhile mathematical tasks.

CONCLUSION

Given the influence tasks have on the learning of mathematics (Stein and Lane 1996), it becomes imperative that teachers understand the relationship between a task and the kind of thinking that task requires of students. "Acquiring the ability to think with precision about mathematical tasks and their use in class can equip teachers with more developed skills in the ways they select, modify, and enact mathematical tasks with their students" (Ball 2000, p. xii). These skills are important ones for teachers to develop, so much so that Ball considers them "a core domain of teachers' work" (p. xii).

Our development of a set of high school–based mathematical tasks to use in a task-sorting activity has supported both preservice and in-service teachers in building knowledge about the activities that form the backbone of any mathematics class. In turn, this new knowledge has enabled teachers to reach a deeper understanding of worthwhile mathematical tasks and the relationship between those tasks and students' learning.

APPENDIX

Eight Tasks from a Task-Sorting Activity

TASK A

Find the smallest positive integer that has exactly 13 factors.

TASK C

State the triangular and unit circle definitions for $\sin\theta$, $\cos\theta$, and $\tan\theta$.

TASK E*

Biologists have determined that the polynomial function

$$p(t) = -0.0001t^3 + 0.002t^2 + 1.5t + 100$$

approximates the population t days later of a certain group of wild turkeys left to reproduce on their own with no predators.

a) Draw a complete graph of the algebraic model $y = p(t)$ of this problem situation.

b) Find the maximum turkey population and when it occurs. Explain how you know this is the maximum population.

c) When will the turkey population be extinct? Explain how you know this date.

Source: Demana, Waits, and Clemens, *Precalculus,* (adapted from p. 184)

TASK F

Find the value of x in the figure below. Write a paragraph that explains how you found x.

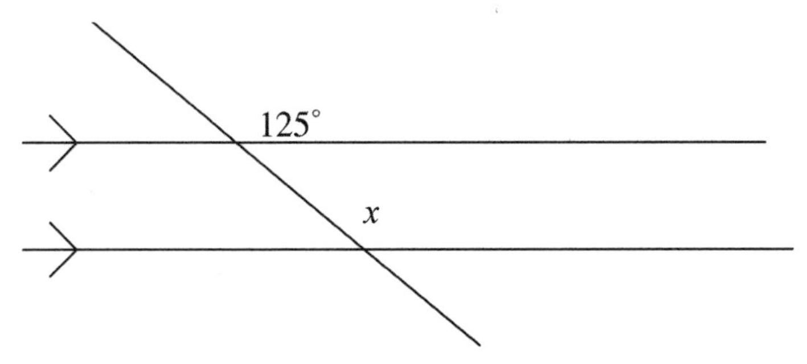

TASK G

A will states that John is to get 3 times as much money as Mary. The total amount they will receive is $11,000.

a) Write a system of equations describing this situation.

b) Solve to find the amounts of money John and Mary get.

TASK J

Postal rates have been figured by the ounce since July 1, 1885. From that date until January 1, 1995, the rates have been as follows:

Nov. 3, 1917	3 cents
July 1, 1919	2 cents
July 6, 1932	3 cents
Aug. 1, 1953	4 cents
Jan. 7, 1963	5 cents
Jan. 7, 1968	6 cents
May 16, 1971	8 cents
March 2, 1974	10 cents
Dec. 31, 1975	13 cents
May 29, 1978	15 cents
March 22, 1981	18 cents
Nov. 1, 1981	20 cents
Feb. 17, 1985	22 cents
April 3, 1988	25 cents
Feb. 3, 1991	29 cents
Jan. 1, 1995	32 cents

On the basis of the data above, predict the cost of mailing a one-ounce, first-class letter in 2010. Explain your reasoning.

TASK K

Match the following rule to its correct name:

1. $a + b = b + a$	a. Identity property for multiplication
2. $(a + b) + c = a + (b + c)$	b. Commutative property of addition
3. $a(b + c) = ab + ac$	c. Transitive property
4. $a + 0 = a$	d. Associative property of addition
5. $a(1) = a$	e. Identity property for addition
6. If $a = b$, and $b = c$, then $a = c$	f. Distributive property

TASK L

Use the table of values below to draw a graph of the function represented. Then use the graph to write the equation of the function. Then use the equation of the function to find $f(5)$, $f(-11)$, and $f(1/2)$. Also use the function to find x if $f(x) = 218$.

x	$f(x)$
1	−1
3	3
0	−3
−2	−7

REFERENCES

Arbaugh, Fran. "Time on Tasks: Influences of a Study Group on High School Mathematics Teachers' Knowledge, Thinking, and Teaching." Ph.D. diss., Indiana University Bloomington, 2000.

Ball, Deborah. "Foreword." In *Implementing Standards-Based Mathematics Instruction: A Casebook for Professional Development,* by Mary K. Stein, Margaret M. Smith, Marjorie A. Henningsen, and Edward A. Silver, pp. ix–xiv. New York: Teachers College Press, 2000.

Doyle, Walter. "Academic Work." *Review of Educational Research* 53 (summer 1983): 159–99.

National Council of Teachers of Mathematics (NCTM). *Professional Standards for Teaching Mathematics.* Reston, Va.: NCTM, 1991.

———. *Principles and Standards for School Mathematics.* Reston, Va.: NCTM, 2000.

Smith, Margaret S., and Mary K. Stein. "Selecting and Creating Mathematical Tasks: From Research to Practice." *Mathematics Teaching in the Middle School* 3 (February 1998): 344–50.

Stein, Mary K., and Suzanne Lane. "Instructional Tasks and the Development of Student Capacity to Think and Reason: An Analysis of the Relationship between Teaching and Learning in a Reform Mathematics Project." *Educational Research and Evaluation* 2 (January 1996): 50–80.

Stein, Mary K., and Margaret S. Smith. "Mathematical Tasks as a Framework for Reflection: From Research to Practice." *Mathematics Teaching in the Middle School* 3 (January 1998): 268–75.

Stein, Mary K., Margaret S. Smith, Marjorie A. Henningsen, and Edward A. Smith. *Implementing Standards-Based Mathematics Instruction: A Casebook for Professional Development.* New York: Teachers College Press, 2000.

4

Mathematical Foundations for a Functions-Based Approach to Algebra

M. Kathleen Heid

Rose Mary Zbiek

Glendon W. Blume

WHEN algebra is taught in the context of constant availability of technological tools, there is a remarkable shift in its content as well as in its pedagogy. In a technological world, instead of focusing on symbolic manipulation, students can turn their attention to symbolic reasoning. Instead of focusing on one symbolic manipulation routine at a time, students can use a variety of representations to investigate algebraic concepts and principles. Instead of spending time on contrived but conveniently illustrative "word problems," students can engage in the initial phases of mathematical modeling.

Technology does more than change the ways students investigate mathematics. It also changes which concepts students study. A functions approach to algebra calls on teachers to work with their students in recognizing function families in a range of applied settings and to reason about functions in a range of representations. In algebra settings that give students access to computer algebra systems, students' attention can be turned to the concepts of variable and function and expanded to the study of families of functions. In such a functions-based algebra (e.g., Fey et al. 1999), the study of families of functions is no longer a list of formulas related to each family. Students focus on concepts such as rate of change, linearity, exponentiality, and "quadraticness." Teachers in these technological classrooms need a deeper knowledge of these concepts and of students' learning of these concepts than those who teach in traditional, nontechnological algebra classrooms.

This work was supported in part by the National Science Foundation under grant no. TPE 91-55313. Any opinions, findings, and conclusions or recommendations expressed in this material are those of the authors and do not necessarily reflect the views of the National Science Foundation.

DEVELOPMENT OF THE FUNCTION CONCEPT

Functions-based approaches to algebra can help students understand the power of knowing how changes in one or more quantities are related to changes in another quantity. Students create and work with functions represented as situations, symbolic or verbal rules, tables of function values, and graphs. Through experience, students learn that different representations give similar information through different forms, and that some representations may be better than others for providing particular information.

To facilitate students' work with functions, teachers need to guide students in the process of learning about a function from all its representations. A starting place for many teachers is prototypical functions and their transformations. For example, $f(x) = x^2$ is a prototype rule for quadratic functions; $g(x) = (x - 3)^2$ and $h(x) = 2x^2 + 6$ represent transformations of that prototype. Using the concept of prototypical function, p, with rule $p(x)$, as an organizing structure, teachers can guide their students in reasoning about graphs of $ap(x)$, $p(bx)$, $p(x - c)$, $p(x) + d$, and eventually composite forms like $ap(b(x - c)) + d$.

Equivalent forms of function rules also take on a new role for teachers and students in a functions-based approach to algebra. Attention turns from producing equivalent forms to interpreting them. Equivalent forms differ in the prominence they give to particular characteristics of a function. For example, $f(x) = (x + 2)(x - 3)$ makes the zeros of f prominent, $f(x) = x^2 - x - 6$ readily displays the value of $f(0)$, and

$$f(x) = \left(x - \frac{1}{2}\right)^2 - \frac{25}{4}$$

highlights the vertex of the parabola that is the graph of f. Not only do teachers need to help students develop an expertise in recognizing the information most accessible in various symbolic forms, but teachers also need to develop their own ability to generate symbolic forms that illustrate given characteristics of families of functions.

Rate of Change

Central to an understanding of differences among families of functions is an understanding of rate of change. Traditionally, investigations of rate of change of functions is a very difficult task, one that has been reserved for an introductory calculus course. Because rate of change is one of the most important characteristics in distinguishing among families of functions, the concept of rate of change is quintessential to a functions approach to algebra. It is our experience in creating and implementing functions-based approaches to algebra and calculus that the conceptual understanding of rate of change, an extension of slope, develops over time and is best developed if approached early and regularly. As early as their first algebra class, students can discuss mathematical concepts related to the concept of rate of change. Teachers need to encourage students to focus on the rate of change instead of focusing only on the amount of change and to articulate the distinction clearly. They might ask students to submit a journal response clarifying the difference through an example from their everyday life.

Students often give curtailed or unclear answers, which disguise any confusion they might have about the difference between change and rate of change. The following discussion mirrors one that occurred in the first few days of a beginning algebra class (Heid et al. 1988). The class was engaged in a discussion of a graph of the average height of females as a function of their age in whole numbers of years. Students had drawn a straight line to describe the data.

Teacher: How did we show the trend in the data ?

Student: Straight lines going up.

Teacher: What would straight lines going up represent?

Student: [*After a few other responses*] A steady increase.

Teacher: [*Teacher draws* , *representing what the student had drawn on paper.*] A steady increase means that something increases at the same rate all the time. Is that true of our data? [*Teacher draws* *to represent the trend in the data.*]

Student: The first part of the curve goes up a different amount than the last part of the curve.

Teacher: What do you mean by "a different amount"?

In this exchange the teacher focused the student's attention on what the graphs represented and what the student meant by "a different amount." The teacher saw this as a teachable moment, one in which she needed to find opportunities for the student to find and compare amount of change and rate of change. The discussion continued with the teacher asking whether someone could describe differences in the way the nonlinear curve changed at different parts of the graph. After a few other answers were offered and explored, one student responded, "At first, it increases a little, and later, it increases a lot." The ensuing discussion involved the teacher facilitating the student's distinguishing between amount of change and rate of change, with a primary feature of the discussion centering on a comparison of the amount of change over different intervals of the same size, using the drawings shown in figure 4.1. Here, the teacher decided to help students focus on the difference between amount of change and rate of change by asking each of them to draw a linear and a nonlinear graph and to identify parts of the graphs that represented equal changes in the input values. The teacher then asked students to decide in groups whether equal changes in input values always give equal changes in output values. She then asked groups to present their arguments. For any two same-sized intervals of input values on a line, the change in output values was the same. For two same-sized intervals of input values on a nonlinear curve, the change in output values could be different. The teacher used the term *rate of change* in a natural setting, and students connected the term to the subtraction of successive quantities in computer-generated tables of values. At first the students talked only about the change in output values, *new y* – *old y*, instead of the needed ratio of the change in output values to the change in input values,

$$\frac{new\, y - old\, y}{new\, x - old\, x}.$$

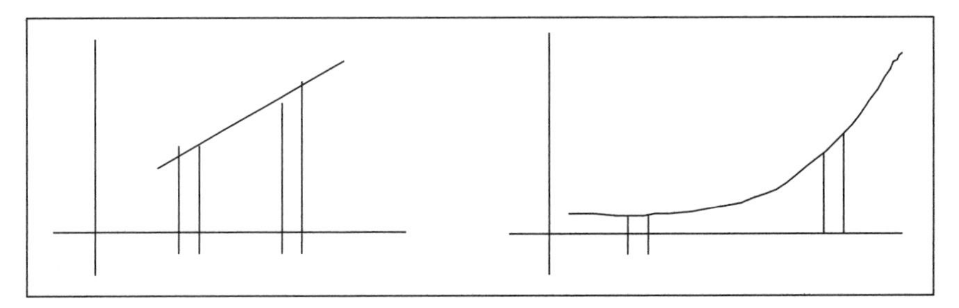

Fig. 4.1. Sketches used by the teacher in comparing change and rate of change

The fact that many of the tables they create or encounter have input values that differ by 1 complicates students' difficulty in distinguishing amount of change from rate of change. To counter the conflict presented by such tables, teachers can point out that the differences of interest are calculated from any two input-output pairs for a function, not just from successive pairs. Moreover, by using and discussing tables with input values that differ by amounts other than 1, the teacher helps the students move beyond this confusion. One way teachers can help their students attend to the difference between change and rate of change is to engage classes in mock debates in which students take opposite sides on the issue of "Change and rate of change? Are they the same or different?" Another way is to construct sets of situations for students in which it matters whether change or rate of change is being discussed. For example, students might be asked to consider the following situation:

Customers at a local gas station noted each time the price of a gallon of gas changed over a 36-day period. They recorded their results in a table like the following:

Day	Price of one gallon
7	\$1.29
14	\$1.43
28	\$1.57
35	\$1.71
42	\$1.85

What was the change in price from day 7 to day 14?
What was the change in price from day 14 to day 28?
What was the rate of change in price?

The follow-up discussion could center on issues like the difference between talking about change and about rate of change (from day 14 to 28, the change in the price of a gallon is 14 cents and the rate of change in the price of a gallon is 1 cent per day), the importance of specifying units when calculating a rate of change (from day 7 to day 14, the rate of change could be \$0.14 per week or \$0.02 per day), and the importance of a constant rate of change in a linear function (this function is not linear, since the rate of change in the price of a gallon is not constant: it is 1 cent per day from day 14 to day 28 and 2 cents per day from day 7 to day 14).

Families of Functions

Functions-based approaches to algebra commonly center on a small set of families of functions—linear, quadratic, and exponential, for example. Any of these families can then be described as a set of functions with a common symbolic form. In the remainder of this article, we will focus on linear functions of the form $f(x) = ax + b$, quadratic functions of the form $h(x) = ax^2 + bx + c$, and exponential functions of the form $g(x) = ca^x$. These particular families of functions are especially useful in describing real-world situations.

Linearity

Although many introductory algebra texts have chapters on linear equations, linear functions are treated differently in many functions-based approaches. Teachers shift students' foci from symbolic routines associated with linear functions to the concept of linearity. They engage students in identifying linear relationships among quantities. Students learn to recognize constant rates of change as being necessary and sufficient to determine linearity and, consequently, to recognize relationships that are not linear. Teachers can engage students in considering linearity in a variety of representational settings. One of our favorite tasks is a card-sorting task (Zbiek 1992). Students are given cards, each with a table, rule, situation, or graph of a function, and then asked to sort those cards according to structure. For example, students are given the twelve cards in figure 4.2, each containing one of the tables, rules, situations, or graphs shown in the figure. Students quickly place h and p in a "linear functions" pile and eventually add b and t to this pile. They may quickly place g in a "quadratic function" pile and then put c, r, and z in a pile of "other things." Students group a and s, then j, and eventually v as constant functions, but they fail to note that these also are linear functions. Giving students the opportunity to sort the cards provides a way to see whether students fail to identify the constant functions as linear functions. Watching students while they perform this sorting task can help teachers notice which representational forms may be more difficult for certain types of functions or for particular students.

In accomplishing tasks like this one, students recognize linearity in graphical, numerical, and symbolic representations:

- Graphically—functions with constant rates of change will graph as straight lines.
- Numerically—output values in a table for a nonconstant linear function with uniformly incrementing input values will either uniformly increase or uniformly decrease.
- Symbolically—linear functions can be expressed in the form $f(x) = ax + b$, where a is the constant rate of change and b is the value of $f(0)$.

Teachers can raise general issues about linearity in a wide range of settings. For example, students may encounter step functions and wonder, because their graphs are composed of line segments that appear to students to form a line, whether these functions are linear. Teachers who are alert to this misconception can engage students in explorations to clarify the issue. One useful exploration involves a postage-rate function that describes the 2002 cost of first-class mail as a function of the weight of a package in ounces. The function is defined for positive input values no greater than 11 (the symbol $\lfloor x \rfloor$, called the "floor function," represents the greatest integer less than or equal to x):

$$p(x) = 23\lfloor x - 1 \rfloor + 37 \text{ if } \lfloor x \rfloor = x$$

and

$$p(x) = 23\lfloor x - 1 \rfloor + 37 \text{ if } \lfloor x \rfloor = x$$

(We chose to use *floor function* rather than *greatest integer function* because, as computer scientists have suggested, the phrase *floor function*, and the related *ceiling function*, is much more suggestive of its meaning to learners. Having the floor function and the ceiling function as fundamental parts of their function repertoire also provides students an advantage in mathematical modeling tasks.) The screens in figures 4.3 and 4.4 show this piecewise-defined function displayed on a symbolic calculator, in a table of values for small numbers of ounces, and as a graph of p.

TABLES

x	$a(x)$	x	$b(x)$	x	$c(x)$
4	3.14	1.0	0.9	1	1
6	3.14	1.1	0.8	2	2
8	3.14	1.2	0.7	3	2
10	3.14	1.3	0.6	4	3
12	3.14	1.4	0.5	5	2
14	3.14	1.5	0.4	6	4
16	3.14	1.6	0.3	7	2
18	3.14	1.7	0.2	8	4
20	3.14	1.8	0.1	9	3
22	3.14	1.9	0.0	10	4

RULES

$g(x) = x^2 + 3x - 7$ $h(x) = 419 - 211x$ $j(x) = \sqrt{1.784}$

SITUATIONS

t Kinko's charges six cents for each plain white copy. A customer's bill for plain white copies is a function of the number of copies he or she has made.

v Every person in your class correctly counts the number of red stripes on the American flag. The number of stripes counted is a function of the age in months of the person who counts them.

z You count the number of people in each car that passes you on an interstate highway. The number of people in a car is a function of the number of cars that have passed you.

GRAPHS

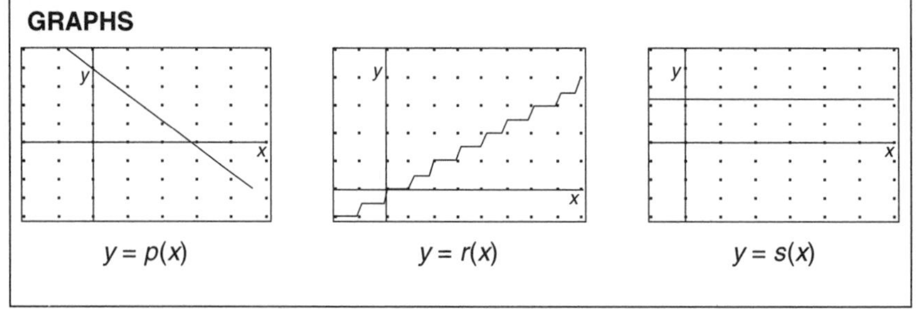

Fig. 4.2. Card-sorting task adapted from Zbiek (1992)

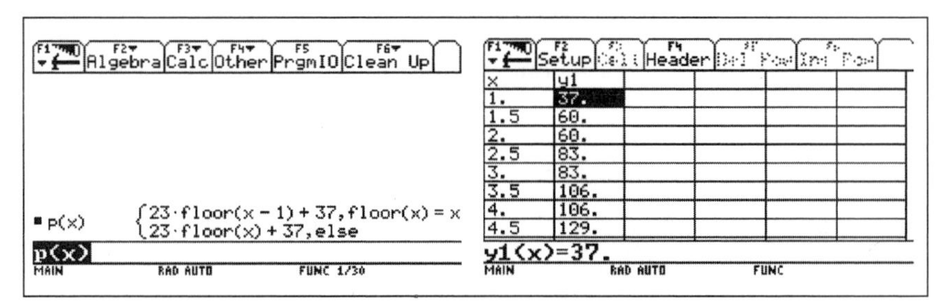

Fig. 4.3. Symbolic rule and table of values for postage-rate function

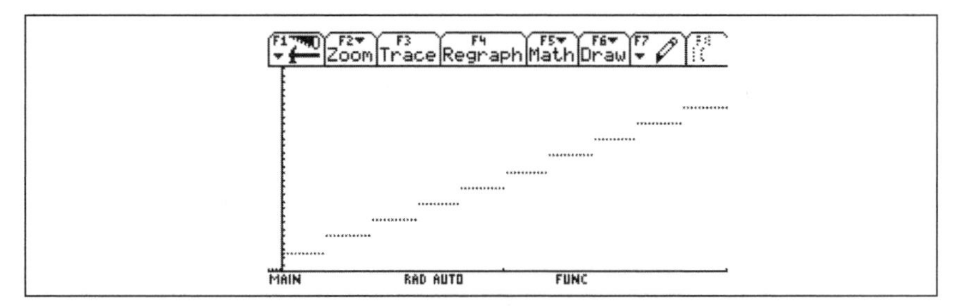

Fig. 4.4. Graph for the postage-rate function

The graph of p is composed of horizontal line segments. To draw attention to rate of change and linearity, the teacher asks students to calculate rate of change for different pairs of output values whose input values differ by the same amount. For example, by computing the rate of change between 2.25 and 1.75 (i.e., $p(2.25) - p(1.75) = 23$) and between 1.75 and 1.25 (i.e., $p(1.75) - p(1.25) = 0$), students see that the rate of change for this function is not constant.

Teachers recognize that linear functions arise any time that a quantity increases at a constant rate. Linear relationships often arise in economic situations because of the common occurrence of a constant unit price. For example, if a car rental agency charges \$63 per day in addition to \$0.25 per mile traveled, then the daily charge to the customer is a linear function c of the number of miles, n, traveled that day with rule $c(n) = 63 + 0.25n$. Another example assumes that 3500 people are expected to attend an event if the admission is free, and the number of people one could expect to attend will drop by 250 for every dollar increase in admission fee. The number expected to attend is a function, n, of the admission fee, p, with rule $n(p) = 3500 - 250p$.

Even if linear functions arise in many places, students often assume their existence even more frequently than is appropriate. These erroneous assumptions are sometimes hidden from all but the most discriminating listener. Although it is sometimes reasonable to linearize in approximating an almost linear situation, students sometimes assume linearity inappropriately without realizing that they are doing so. Imagine the following scenario. The students have studied exponential growth in the past and are now asked to estimate the population at the end of week 2 if the population at the end of week 1 was two thousand and the population at the end of week 3 was eight thousand.

Student: Five thousand! The population will be five thousand.

Teacher: How did you figure that out?

Student: Well, five thousand is halfway between two thousand and eight thousand.

Teacher: Can you draw a graph to show what you mean?
[*Student draws a graph of* .]

Teacher: Now show me a graph that illustrates the data you used.
[*The student produces something like the following.*]

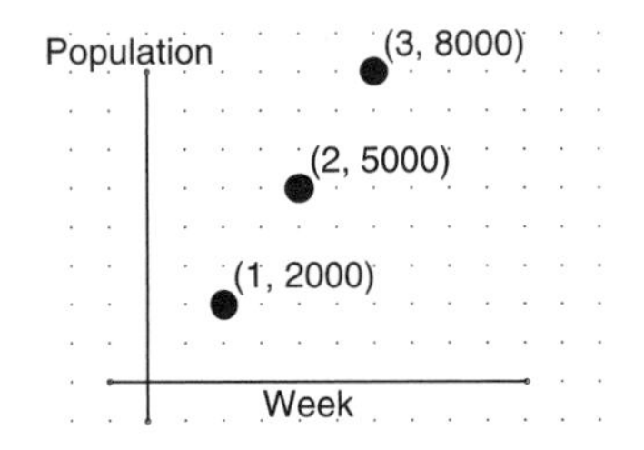

Teacher: Are you satisfied with your two drawings?

Student: They look kind of different.

Teacher: Can you explain why?

Student: I'm confused: the curved graph says that it should be less than five thousand!

At this point, the student seemed to be experiencing cognitive conflict and needed to resolve the conflict. The teacher would continue the discussion with the student until he saw the linear model implicit in his interpolation process. The teacher could, for example, have asked the student to construct two tables of values, one for a linear function of the student's choice and one for a function of the form $f(x) = ca^x$ ($a > 0$, $c > 0$), using 2, 4, and 6 as input values. The teacher might next have asked the student to predict and then calculate the output values for input values of 1, 3, and 5. Subsequent discussion would center on the differences in the functions (one is linear and one is not) and the rationale behind linear interpolation.

Exponentiality

Along with linear functions, exponential functions of the form $f(x) = ca^x$ ($a > 0$, $c > 0$) represent a growth pattern that is common in students' everyday lives. However, students confuse the multiplicative growth pattern of exponential functions with the additive growth pattern of linear functions. Both growth patterns can be characterized recursively, since the "next" value is always determined by the "current" value. If f is a linear function, then the value of $f(n + 1)$ can be calculated by adding a constant to the value of $f(n)$, regardless of the value of n. Similarly, for exponential functions of the form $f(x) = ca^x$ $(a > 0, c > 0)$, the value of $f(n + 1)$ can be calculated by multiplying the value of $f(n)$ by a constant (in this instance, a), regardless of the value of n. Teachers can capitalize on the iterative nature of exponential functions by focusing students' attention on technology-generated tables of values. It is common for students to examine the entries in a table and remark on "how to get the next one." Teachers can build on this tendency by engaging students in constructing iterative function rules from tables. Figure 4.5 gives an example of such an exercise.

The exponential functions students study in functions-based, technology-intensive curricula share some characteristics with some of the other functions they study. These functions are monotonic, like linear functions. They increase (or decrease) at a nonconstant rate, like quadratic functions. Such similarities force teachers to be alert for possible misconceptions. To the naive eye, an exponential function may look like "half a parabola." Students who make such observations may be thinking about functions only in terms of general shape instead of more precise properties. It is important for teachers not to let misconceptions like this go unchecked and to emphasize to students the importance of reasoning from definitions.

Students should recognize exponential situations in everyday living. Students learn that any situation for which growth is describable by a constant percent increase or decrease in amount can be modeled by an exponential function of the form $f(x) = ca^x$ $(a > 0, c > 0)$. For example, exponential functions can be used to approximate the value of a car or house after a given number of years assuming that the value of the car or house increases or decreases at a rate proportional to its current value (*new* $y = a \cdot$ *old* y). Exponential functions can be used to describe the number of bacteria in a culture after a specified amount of time, given that the number of new bacteria is a constant percent of the number of bacteria in the culture at a given time. Exponential functions can be used to describe the amount of radioactive material that remains after a given amount of time if it is assumed that the amount of decay is directly proportional to the amount of radioactive material present at a given time.

Teachers can use the study of exponential relationships as an occasion to engage students in intensive mathematical connections. It is common knowledge that in the instance of linear growth like $y = 1.05 \cdot x$, the value of y is directly proportional to the value of x. In the case of an exponential

Public health departments monitor the cleanliness of restaurants, grocery stores, swimming pools, and other public facilities, because harmful bacteria can grow quickly in untreated conditions.

For instance, data in the following table show a typical pattern of bacterial growth in water of a swimming pool if no filtration or chlorine is used.

Time (days)	**Number of bacteria per cubic centimeter**
0	1000
1	2000
2	4000
3	8000

Copy and complete the table for input values through 10 in a pattern that matches the entries in the first four rows.

The pattern in the table is easy to extend and describe. From each day to the next, the bacteria population density doubles:

$$2000 = 1000 \times 2$$
$$4000 = 2000 \times 2$$
$$8000 = 4000 \times 2,$$

and so on. If d represents the time in days and $N(d)$ represents the number of bacteria per cubic centimeter on day d, then the above pattern can be written as follows:

$$N(2) = N(1) \times 2$$
$$N(3) = N(2) \times 2$$
$$N(4) = N(3) \times 2,$$

and so on. This pattern can be summarized by the rule

$$N(d + 1) = N(d) \times 2.$$

This means that the number on any given day equals twice the number on the previous day.

Fig. 4.5. An example explaining the construction of iterative function rules from tables (Fey et al. 1999, pp. 239–40)

curve like $y = 20 \cdot 1.05^x$ the amount of growth in y is directly proportional to the value of y at a given time. That is, for any two pairs of (x, y) values that satisfy the equation $y = 20 \cdot 1.05^x$, the value of

$$\frac{\Delta y}{y},$$

or alternatively

$$\frac{new\ y - old\ y}{old\ y},$$

is constant. For $y = 20 \cdot 1.05^x$,

$$\frac{\Delta y}{y} = \frac{new\ y - old\ y}{old\ y} = 0.05,$$

as the calculations in figure 4.6 suggest.

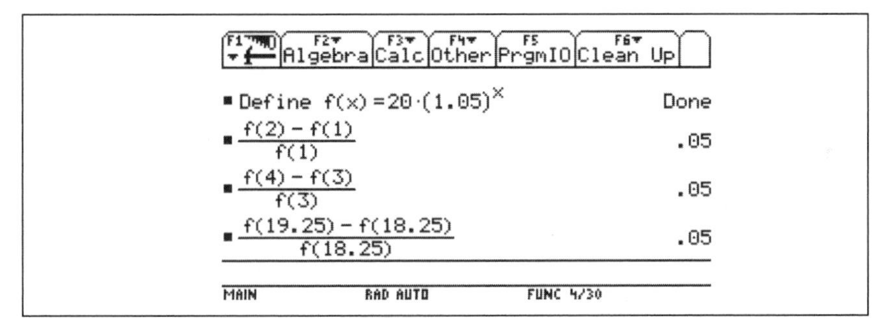

Fig. 4.6. Illustration of the fact that $\frac{new\ y - old\ y}{old\ y}$ is a constant in the function $y = 20 \cdot 1.05^x$

This characterization of the exponential highlights the iterative nature of exponential functions. Since

$$\frac{new\ y - old\ y}{old\ y} = 0.05,$$

it follows that $new\ y - old\ y = (0.05) \cdot old\ y$. Working from this new equation leads to a recursive formula for this exponential function:

$$new\ y = old\ y + (new\ y - old\ y)$$
$$new\ y = old\ y + (0.05) \cdot (old\ y)$$
$$new\ y = (1.05) \cdot (old\ y)$$

In recursive notation, the last of these equations becomes $u(n) = 1.05 \cdot u(n-1)$ and $u(0) = 20$.

Teachers can use the study of exponential functions to help students to see situated examples of negative and fractional exponents and to consider the meaning of those exponents graphically and numerically. Consider the following task: Suppose $g(d) = 2^d$. The value of $g(d)$ for $d = 0.5$ is $\sqrt{2}$. Use successively refined graphs or tables to find the approximate decimal value for $\sqrt{2}$. Teachers also can help eliminate the mystery of negative and decimal exponents when they help students view them as values of a function that describes a real-world context. For example, suppose the number of bacteria in a pool at 8:00 a.m. on Monday was 1500 and $N(d)$ represents the bacterial population in a pool d days after that with $N(d) = 1500 \cdot 2^d$. Then $N(-4)$ represents the number of bacteria in the pool four days earlier (the preceding Thursday), and $N(0.5)$ represents the number of bacteria in the pool a half day later at 8:00 p.m. on Monday. Because students learn about exponential functions in context, there is little room for some of the usual confusions (e.g., they seem not to confuse the meanings of 2^x and x^2).

Quadraticness

The study of quadratic functions can be students' first encounter with nonconstant rates of change. It is not uncommon for students to conclude that any function whose graph is not a straight line must be quadratic. Attending to this potential misconception, teachers include nonlinear and nonquadratic examples early in students' work with functions. As students' repertoires of function families expand, teachers continually have students compare and contrast new families with the more familiar ones. For example, students can distinguish exponential functions from quadratic functions by noting that quadratic functions have minimum or maximum values.

Quadratic functions do not seem to arise in real-world situations as obviously as linear or exponential functions do. In locating or creating additional quadratic functions, teachers sometimes find it helpful to think of the quadratic as a product of two linear quantities, each dependent on the same quantity. For example, the area of a rectangle is a product of length and width. If the length and the width are linear functions of x, then the area of the rectangle is a quadratic function of x. Similarly, a revenue function is expressible as the product of the number of items that are sold and the price per item. When the number of items that are sold is a linear function of price, then the revenue function is a quadratic function of price.

CONCLUSION

As teachers become involved in implementing a functions-based approach to algebra, they become increasingly challenged to meet the content-knowledge demands of the approach. They need to think about familiar ideas in unfamiliar ways, and they need to think more broadly about the concepts that in the past have been reduced to routine. They need to be prepared to work with their students in applying function families to a range of settings and to reason about functions using a variety of representations. Concepts such as rate of change, the nature of linearity, contextual meaning of rational and negative exponents, quadraticness, and exponentiality become newly important in functions-based approaches to algebra. Precision in thinking about related ideas (such as amount of change versus rate of change and proportionality of amounts versus proportionality of rates) becomes increasingly important as students' mathematical work is organized around the concept of function. Some ways of thinking (e.g., reasoning from definitions, thinking about linear and exponential functions iteratively) take on new importance in technological settings. In this article, we have discussed essential features of relatively familiar families of functions and have provided ways to think about conceptual and pedagogical challenges that arise as teachers and students engage in a functions-based approach to algebra.

REFERENCES

Fey, James T., M. Kathleen Heid, Richard A. Good, Charlene Sheets, Glendon W. Blume, and Rose Mary Zbiek. *Concepts in Algebra: A Technological Approach.* Chicago: Everyday Learning, 1999.

Heid, M. Kathleen, Charlene Sheets, Mary Ann Matras, and James Menasian. "Classroom and Computer Lab Interaction in a Computer-Intensive Algebra Curriculum." Paper presented at the annual meeting of the American Educational Research Association, New Orleans, April 1988.

Zbiek, Rose Mary. "Understanding of Function, Proof, and Mathematical Modeling in the Presence of Mathematical Computing Tools: Prospective Secondary School Mathematics Teachers and Their Strategies and Connections." Ph.D. diss., Pennsylvania State University, 1992.

5

Language Pitfalls and Pathways to Mathematics

Carne Barnett-Clarke

Alma Ramirez

THE meaning we construct for a mathematical idea is closely entwined with the way we learn to use language to reason about the idea and to translate among words, symbols, and meanings of that idea (Cobb, Yackel, and McClain 2000; Pimm 1995). For example, if a student wrote 41 for the sum of 5 + 9, a teacher might think that the child was guessing or was simply writing their numbers backward. Yet, when asked orally, the same child might reply that the sum is fourteen. After all, doesn't it makes sense that if you write the *four* first for *forty-two* that you would write the *four* first for *fourteen*?

Although this example may seem simplistic at first, consider the consequences for thousands of students. First, the inconsistencies in how students say and write 14 may have an impact on how they understand and represent place-value concepts. Also, language pitfalls such as this may easily lead to an inaccurate, or perhaps incomplete, assessment of what students know and can do. Furthermore, students may be criticized unjustly for not paying attention or not trying hard enough when they make what appear to be careless errors.

As teachers, we must learn to carefully choose the language pathways that support mathematical understanding, and simultaneously we must be alert for language pitfalls that contribute to misunderstandings of mathematical ideas. More specifically, we must learn how to invite, support, and model thoughtful explanation, evaluation, and revision of mathematical ideas using correct mathematical terms and symbols. In addition, we need to build a repertoire of the common ways that language can be ambiguous or misleading to students so that we can anticipate, reveal, and untangle misunderstandings before they become stubbornly embedded in students' thinking. Furthermore, researchers in the field of language acquisition and mathematics point out that these confusions are significantly compounded for students who are acquiring English as a second language (Cuevas 1991; Khisty 2002; Olivares 1996). These students must contend with many more language layers than their native English-speaking peers and thus require additional scaffolds to acquire mathematical linguistic competence in English.

THREE ASPECTS OF LANGUAGE IN MATHEMATICS CLASSROOMS

There are many dimensions to the influence of language in teaching and learning mathematics. Three aspects are the symbols and vocabulary inherent in mathematics as a discipline, students' internal chatter, and classroom interactions through discourse.

Mathematical Symbols and Vocabulary

One aspect of language that is crucial to mathematics learning pertains to the specialized terminology, symbols, and syntax used to express mathematical ideas. In fact, the vocabulary and symbols of mathematics make it similar to learning a foreign language. Not only do students need explicit instruction to read and write mathematical symbols and words, they also need to learn how to express mathematical ideas orally and with written symbols. Students must also develop meaning for individual symbols such as the equals sign, as well as for mathematical expressions and sentences such as $3x + 4y$.

Many times the use of the symbols themselves is confusing. For example, letters can represent abbreviations such as in 15 m, where "m" represents a meter, a unit of measure. Letters can also represent variables such as in $5w = m$ where w represents a number of women and m represents a number of men. In other words, $5w = m$ is read as "5 times the number of women is equal to the number of men," not "5 women equals 1 man." Students learning algebra may be confused if this distinction is not pointed out explicitly.

Another difficulty is that vocabulary, symbols, and syntax are often different when used in a mathematical context rather than in an everyday-English context. For example, consider the table where you eat breakfast versus the table that shows the values of the variables in an equation. There is a high risk of misunderstanding if this specialized meaning is taken for granted and not specifically pointed out to students. Research has shown that this is especially true for the students whose primary language is not English. These students not only have to understand the meaning of a word like *table* in everyday English, but they also must refer back to their primary language for its meaning, gain fluency in using the word, and then repeat the whole process again when they learn the word in mathematics. Imagine if no one pointed out the difference between the colloquial usage and the mathematical usage of the word.

Classroom implications

- To help students build their mathematical vocabulary, invite them to read mathematical words, symbols, expressions, and equations aloud in chorus, have them read to a partner, or ask them to read along silently as you or another student reads aloud.
- Provide many opportunities for students to translate mathematical words, representations, or problems into symbols and vice versa.
- Point out not only what terms mean but also how they are used in particular instances. This is especially important for students who are learning English as a second language. For example, students need to know not only what a term like *half* means but also what it means when used in an expression like *half as much.*

Internal Chatter

Take a moment to think about how to solve the number sentence $0.5 \times \_\_\_\_ = 30$. In solving this problem, you must speak to yourself as you think, either aloud or mentally, in what Bickmore-Brand (1990) calls internal chatter. For example, you might say to yourself, "Five-tenths of some number is the same as 30. Five-tenths is half, so half of some number is 30. The number must be 60. That makes sense because half of 60 is 30." Teachers help students by modeling and mentoring this kind of internal chatter, or reflective talk. The goal is for students eventually to adopt the same kind of thinking, reasoning, and self-questioning as modeled and mentored by their teacher.

Internal chatter also supports the thinking process necessary for solving problems. Consider the following problem:

> Jesse saved 80 cents in nickels.
> How many nickels does he have?

One way to approach the problem is to talk internally about what it means, "Jesse saved some nickels that are worth 80 cents. I want to find out how many nickels he saved. I know each nickel is worth 5 cents, so I need to find out how many fives are in 80." Once the problem is clear, a solution process is more likely to emerge; however, researchers point out that the linguistic features of problems such as this one contribute to its difficulty level (Spanos et al. 1988). Notice, for example, that common language cues such as *all together, how many left,* or *shared among* are not available to signal which operation to use. Also, students have to understand what it means to say "80 cents in nickels." Words such as *in, of, by, from,* or *into* often have special meanings in mathematics and may be difficult to interpret, especially for English-language learners.

For students whose first language is not English, internal chatter is likely to occur in the primary language first or in a combination of the primary language and English. The complexity of this problem is also increased, since only one number, 80, is obvious in the problem. Students must interpret the meaning of a nickel as 5 cents, and then restate the problem using this information before it can be solved.

In a language-impoverished mathematics lesson, either there is little talk during instruction or the discourse focuses primarily on the explanations of the steps, key words, or simple definitions. In such an environment, students do not develop the kind of rich internal chatter that helps them make sense of problems. Instead, their reflective talk focuses on remembering and performing the steps in the pattern.

For example, when they see 2/3 + 2/3 = ____, they try to remember a pattern for doing this kind of problem. Their internal chatter might go something like this: "It's a plus problem. Two and 2 is 4, and 3 and 3 is 6. That's 4/6. I need to reduce. Two into 4 is 2, and 2 into 6 is 3. The answer is 2/3." They don't stop to ask themselves whether it makes sense to add 2/3 and 2/3 and get 2/3! They haven't developed a little voice inside that prompts them to think about the meaning of the problem or to question and monitor their thinking as they go along. They just do steps as they can remember them.

Classroom implications

- Solve problems or equations aloud with students, or have them work with a partner to help them develop the internal chatter to support their thinking.
- Don't overlook what might seem obvious: students often do not know what words to use for symbols such as a blank, the equals sign, or parentheses in an equation.
- Allow students who are learning English to work in small groups with others who share the same primary language. This allows them to work together to verbalize the mathematical problems and solution processes in their stronger language before tackling them in another language.
- Assist students who are discussing in small groups by listening closely to their ideas, asking clarifying questions, modeling, paraphrasing, and writing down their ideas to provide them with a record of their mathematical talk.

Classroom Discourse

Simply put, an important point of classroom discourse is to develop students' understanding of key ideas. Discourse is multifaceted. It involves both receptive communication (listening, reading, interpreting diagrams and pictures, and interpreting others' actions), as well as expressive communication (speaking, writing, drawing, performing, demonstrating, and imagining). Discourse is most effective if both students and their teacher are involved in both expressive and receptive communication.

There are many ways to structure classroom discourse. For example, in a "community of learners" model, a teacher might pose a problem and then have several students share their solutions. The purpose of the discourse is to build an understanding of the mathematics by offering a forum for unraveling solution processes, making mathematical arguments and proofs, and uncovering and analyzing pitfalls in mathematical ideas. The teacher's role is to move the mathematics along while affording students opportunities to offer solutions, make claims, answer questions, and provide explanations.

Another structure for classroom discourse is for a teacher to model how to solve a problem by thinking aloud using appropriate language, methods of reasoning, and metacognitive processes to decide the next steps or evaluate thinking for pitfalls (Wenger 1999; Schoenfeld 1992). Throughout the solution process, the teacher involves students in dialogue about the thinking process to foster their understanding and encourage their cognitive engagement. The eventual goal is to reverse the roles, where students think aloud as they solve problems while the teacher offers helpful prompts and questions to focus attention on particular ideas. By first modeling and then mentoring learning in this way, the teacher provides a scaffold, or temporary support, that is eventually no longer needed (Vygotsky 1978).

Notice that in both of these discourse structures there is an emphasis on receptive and expressive roles for the students and also for their teacher. What may not be explicitly considered, however, is how the teacher needs to pay deliberate attention to having students read, write, and say mathematical words and symbols as part of any discourse process. For example, it is helpful for the teacher to point out and discuss the meanings of new words or symbols and to write them on the board. In addition to modeling the correct use of mathematical language, the teacher must also support the correct use by students.

For some students, especially those learning English, aural word recognition may not be accurate or quick enough for the students to be able to follow the discussion. If talk is chaotic or unclear, the benefits of discourse for these students are significantly reduced. It will help if the teacher uses voice inflection, modified speech, a reduced pace, pauses, and body language to stress particular points or to indicate a transition from one part of the discussion to another (Khisty 2002).

Classroom implications

- During classroom discourse, ask students questions to elicit their understanding and ideas and encourage them to raise their own questions. They should use both expressive and receptive communication.

- Encourage students to use their primary language during small-group discussions, especially when a concept is cognitively and linguistically demanding.
- During the discussion, provide linguistic supports such as diagrams and visual aids. Clarify and rephrase as needed.
- Create a written record of the important points made during the discussion so that students, especially those learning English, can follow along more easily.

LURKING LANGUAGE PITFALLS

Several language pitfalls occur as part of mathematical teaching and learning. Pitfalls are prevalent misconceptions or inaccuracies that have logical and intuitive roots and are resistant to change. The pitfalls we describe are most often related to content, not just mistakes in vocabulary use.

Peculiarities of Mathematical Words, Symbols, and Syntax

Mathematical language contains many inconsistencies that easily confuse students. These inconsistencies are often so much a part of everyday use that adults often don't realize the potential pitfalls that lurk in how students interpret what is being said or written. For example, teachers must help students, especially English-language learners, distinguish between homonyms, such as *sum* and *some* or *sign* and *sine*, as well as understand the interchangeability of synonyms such as *add, plus, combine, increase by,* and *join* (Olivares 1996; Thompson and Rubenstein 2000). We can avoid confusion by pointing out the distinctions or shared meanings explicitly by saying, "Another word that we use to mean *add* is *combine*."

Another difficulty is that the very nature of mathematical language as it is used in textbooks and word problems is different from ordinary English. For one thing, the syntax of word problems is unfamiliar to students and difficult to unpack and interpret. In addition, the meaning of the mathematical text is based on commonly accepted conventions that must be learned. For example, in solving word problems, students need to learn that "the difference between 9 and 14" is usually assumed to mean that you are to subtract 9 from 14. Notice that the difference between 9 and 14 could also be that 14 has 2 digits and 9 has 1 digit. Although this is also a correct interpretation, students need to learn when this interpretation is acceptable and when it is not. This is part of learning the conventional use of mathematical language. The following examples illustrate some of the ways mathematical words and symbols can be confusing.

Why don't decimals have oneths?

Students might expect symmetry around the decimal point when writing numbers in decimal form. Yet, to the left of the decimal point are the ones place, tens place, and hundreds place. To the right of the decimal point are the tenths place, the hundredths place, and the thousandths place. Students might wonder, "Where is the oneths place?"

In an effort to help students see the relationships of whole numbers and decimals, a teacher might say, "Decimals are just like whole numbers." Yet, students find the symbols for writing decimals unexpectedly different from the symbols for whole numbers. Both the similarities and the differences in saying, writing, and reading decimals need to be made clear to students.

Oral and Symbolic Confusions

There are many irregularities between the oral names for symbols and their written form that confuse students. There may also be incongruence between what the teacher means and what the students hear.

> Why might it make sense to write 50032 for five hundred thirty-two?
>
> Why don't you write 3/5 as the decimal 3.5?
>
> Why do you say "*x* over 3" for $x/3$ instead of "*x* thirds"?

When a teacher is talking about a whole, could the student be thinking a "hole"?

Classroom implications

- Give cues about the importance of terms and expressions using body language, voice tone, and inflection. Write mathematical words and symbols on the board and call attention to how they are spelled and pronounced.
- Have students read mathematical words, symbols, expressions, and equations aloud. Point out irregularities in the way terms are written, symbolized, or verbalized.

Oversimplifying with Key Words and Definitions

In an effort to make learning a concept simple, sometimes we oversimplify the meaning of mathematical words or symbols in our explanations. For example, when we tie a concept to a key word, restricted label, or overly simple definition, we risk limiting students' abilities to think about the concept in more than one way or to apply the concept in situations that are unfamiliar.

We must be particularly careful when emphasizing key words, especially with second-language learners. These students are particularly susceptible to overreliance on key words because their reading comprehension in English is limited. An overemphasis on this strategy may prompt them to decode word problems and perform computation on the numerical values they see instead of trying to read the problems for meaning and deciphering contextual and semantic structures.

The following examples point out some typical pitfalls related to oversimplification.

Word problems: Key words and direct translations

One way we try to simplify problem solving for students is to have them look for key words that signify particular operations. Another way is to teach students to translate the words in the problem directly to an equation. Although both these strategies can be helpful, they also have limitations. Consider the following problem:

> Judy ate 3 times as many cookies as Benito.
>
> Judy ate 6 cookies. How many did Benito eat?

Students who try to use a strategy of direct translation for this problem run into difficulty because they pick out the number *3*, the key word *times*, then the number *6*, in that order. So it would not be surprising if students incorrectly wrote the equation $3 \times 6 =$ ____ instead of $3 \times$ ____ $= 6$. Notice that although this problem may seem easy on the surface, students must make sense of the complex comparative phrase "3 times as many."

Learning to pay attention to cues that commonly signify a particular operation is a helpful aid; however, many word problems contain misleading language cues. A key to avoiding pitfalls is to encourage students to make sense of the problem by restating it in their own words, drawing a picture, acting it out, or using a model.

Subtraction means *take-away*

Consider the typical internal chatter a student might use to solve number sentences such as $15 - 12 =$ ____.

> Count out 15, then take away 12. Then count how many are left and write the number in the blank.

The "take away" interpretation of this number sentence is correct, yet there are many ways to articulate the meaning of this number sentence. For example, the problem could be verbalized as "What is the difference between 15 and 12?" or "12 and how many more make 15?" Students who learn to say the problem to themselves in flexible ways can choose a more efficient method to solve the problem.

Classroom implications

- Deliberately develop students' flexibility to verbalize symbols and text to describe and communicate mathematical ideas.
- Use caution when interchanging different words or symbols for the same idea. Explicitly point out to children that there are different ways of communicating the same idea.
- Give students opportunities to talk through problems as a class or with a partner to help them develop skills to make sense of the words. Point out the dangers of relying only on clues such as key words or computing blindly.
- Help students learn how to draw and label diagrams or to use materials to help them represent problems.

Imprecise or Incomplete Explanations and Questions

One of the ways teachers inadvertently lead students to develop misconceptions is by providing imprecise or incomplete explanations or questions. If we model imprecise language or give vague explanations, students will also have difficulty using language in their own explanations or internal talk. In a study of algebra notation, MacGregor and Stacey (1997) found that students sometimes misunderstand what their teachers mean when they say "x without a coefficient means $1x$" (p. 11). Students interpret x as 1, rather than $1x$. Consider the following problem:

> A white rose stem is x units long and a red rose stem is 10 units longer. What is the length of the red rose stem?

When given this problem, students may say the red rose stem is 11 units long, incorrectly interpreting $x + 10$ as $1 + 10$, or 11. The following examples also illustrate how students may miss the point if explanations are vague or not crafted carefully.

How many tens?

How many tens are in 320? A teacher may ask this question, thinking only of the tens place, and expect her students to respond, "Two." If so, she should have asked, "How many tens are in the tens place?" Think about the confusion that this might cause for students.

Adding zeros

Do you just add a zero to 0.45 to make it 0.450, or do you annex a zero? When we say we add a zero, it is possible that students think of adding such as $0.45 + 0$. Even if we say that we annex a zero, don't students need a more complete explanation to understand why 0.45 is equal to 0.450? We also need to be explicit about where the zero is placed. Some students may think that "zeros don't make a difference" and incorrectly conclude that 0.45 = 0.045.

Classroom implications

- Model precise and correct mathematical explanations and reasoning for students. Terms that are used loosely or incorrectly are often the source for students' errors and misunderstandings. If students have difficultly, investigate whether or not the difficulty might be language-based.

Implications for Teachers' Professional Development

The impact of language on mathematics teaching and learning is graphically portrayed in the earlier part of this article; however, where do teachers get a place to wrestle with what to say and do so that children walk away with an understanding of mathematics that has integrity and is resistant to pitfalls? How do they learn about the language-dependant cues and miscues that confuse students, and where do they get a chance to unpack them? How do they learn how to assess what students are saying and how to use this information to promote further learning?

Currently, teachers have few opportunities to reflect on the language pitfalls that impede mathematical understanding or to walk through what they might say to help students understand a complex mathematical idea. Professional development strategies, such as case discussions, the analysis of students' work, and lesson study (Loucks-Horsley 1998), offer natural settings for teachers to have reflective discourse about language issues.

Teachers participating in the Mathematics Case Methods Project (Barnett and Ramirez 1996) were encouraged to analyze the impact of language on their students' mathematical understanding as part of their professional growth. In this project, teachers read and discussed cases of mathematics instruction, paying explicit attention to the impact of language on teaching and learning. They combed a case to find ways that language-dependent cues or miscues might have been confusing to students. They also reflected on how they might solve a problem in the case to consider how language impeded or facilitated learning. Teachers also analyzed what students and the teacher in the case said or wrote to determine how language could be a source of errors or misunderstandings, especially with second-language learners.

Given the complexity of how language intertwines with mathematical understanding, it is imperative that professional development includes opportunities for teachers to think through language issues. We are suggesting that language components could easily be incorporated into different professional development settings. Doing so would help teachers attend to how language impacts students' learning and become more deliberate in their mathematics instruction.

REFERENCES

Barnett, Carne, and Alma Ramirez. "Fostering Critical Analysis and Reflection through Mathematics Case Discussions." In *The Case for Education: Contemporary Approaches for Using Case Methods,* edited by Joel Colbert, Peter Desberg, and Kimberly Trimble, pp. 1–13. Boston: Allyn & Bacon, 1996.

Bickmore-Brand, Jennie, ed. *Language in Mathematics.* Portsmouth, N.H.: Heinemann, 1990.

Cobb, Paul, Erna Yackel, and Kay McClain, eds. *Symbolizing and Communicating in Mathematics Classrooms: Perspective on Discourses, Tools, and Instructional Design.* Mahwah, N.J.: Lawrence Erlbaum Associates, 2000.

Cuevas, Gilbert J. "Developing Communication Skills in Mathematics for Students with Limited English Proficiency." *Mathematics Teacher* 84 (March 1991): 186–89.

Khisty, Lena Licón. "Mathematics Learning and the Latino Student: Suggestions from Research for Classroom Practice." *Teaching Children Mathematics* 9 (September 2002): 32–35.

Loucks-Horsley, Susan. "Effective Professional Development for Teachers of Mathematics." *Ideas That Work: Mathematics Professional Development* 15 (1998): 2–9.

MacGregor, Mollie, and Kaye Stacey. "Students' Understanding of Algebraic Notation: 11–16." *Educational Studies in Mathematics* 33, no. 1 (1997): 1–19.

Olivares, Rafael. "Communication in Mathematics for Students with Limited English Proficiency." In *Communication in Mathematics, K–12 and Beyond,* 1996 Yearbook of the National Council of Teachers of Mathematics (NCTM), edited by Portia Elliott, pp. 219–29. Reston, Va.: NCTM, 1996.

Pimm, David. *Symbols and Meanings in School Mathematics.* London and New York: Routledge, 1995.

Schoenfeld, Alan. "Learning to Think Mathematically: Problem Solving, Metacognition, and Sense Making in Mathematics." In *Handbook of Research on Mathematics Teaching and Learning,* edited by Douglas A. Grouws, pp. 334–70. Reston, Va.: National Council of Teachers of Mathematics, 1992.

Spanos, George, Nancy Rhodes, Theresa Corasaniti, Dale Crandall, and JoAnn Crandall. "Linguistic Features of Mathematical Problem Solving: Insights and Applications." In *Linguistic and Cultural Influence on Learning Mathematics,* edited by Rodney R. Cocking and Jose P. Mestre, pp. 221–40. Hillsdale, N.J.: Lawrence Erlbaum Associates, 1988.

Thompson, Denisse R., and Rheta N. Rubenstein. "Learning Mathematics Vocabulary: Potential Pitfalls and Instructional Strategies." *Mathematics Teacher* 93 (October 2000): 568–73.

Vygotsky, Lev S. *Mind in Society: The Development of Higher Psychological Processes.* Cambridge, Mass.: Harvard University Press, 1978.

Wenger, Etienne. *Communities of Practice: Learning, Meaning and Identity.* Cambridge: Cambridge University Press, 1999.

6

Mathematics Teaching in Grades K–2

Painting a Picture of Challenging, Supportive, and Effective Classrooms

Barbara Clarke

Doug Clarke

MOST of us can recall teachers who made mathematics "come alive" during our time at school—teachers who challenged and yet supported us in our learning, teachers who seemed to know what we needed to learn and the best ways to enhance that learning. But what specifically do effective mathematics teachers do? In this article, we shall describe a classroom research study of effective teachers of kindergarten through grade 2 in Australia that may furnish some pieces of the puzzle. Effective teachers come in many shapes and sizes, but there may be important commonalities as well as differences that can be shared. Our belief is that these findings offer a basis for reflection on good practice, at both preservice and in-service levels, and there is strong research evidence that reforms of the kind advocated by the National Council of Teachers of Mathematics (NCTM) *Principles and Standards for School Mathematics* (2000) are moving in the right direction.

BACKGROUND

For many years, researchers have sought to describe teacher behaviors that correlate positively with growth in students' achievement. In the 1970s and 1980s, the so-called process-product research was conducted (see, e.g., Brophy and Good 1986). In these studies, the dependent variable was perfor-

We wish to acknowledge the insights reflected in this paper of our colleagues in the research team (Jill Cheeseman, Donna Gronn, Ann Gervasoni, Marj Horne, Andrea McDonough, Anne Roche, Glenn Rowley, and Peter Sullivan), Pam Hammond from the Department of Education and Training (Victoria), and the teachers and principals in participating schools in the Early Numeracy Research Project.

mance on standardized achievement tests. Most of these studies occurred in classrooms where the curriculum and classroom organization were designed to support the teacher in *direct instruction*. Not surprisingly, therefore, most of the effective teacher variables involved teacher behaviors associated with this style of teaching, including the clarity of presentation, teacher wait time, and whole-class questioning strategies.

This research is somewhat limited in what it offers to teachers who are seeking to operate in a classroom where meaning and understanding are the focus, although clarity of explanation, appropriate wait time, and questioning strategies are, of course, still desirable. As will be discussed, an important difference between these studies and the present study was the quality of the assessment information that emerges from a one-on-one interview, particularly with young children.

Through observation and videotape, Stigler and Stevenson (1991) conducted a study of 120 classrooms in Taipei (Taiwan), Sendai (Japan), and Minneapolis (United States). They characterized the classrooms in Japan and Taiwan as follows (p. 14):

> Coherent lessons ... are presented in a thoughtful, relaxed, and nonauthoritarian manner. Teachers frequently rely on students as sources of information. Lessons are oriented towards problem solving rather than rote mastery of facts and procedures and utilize many different types of representational materials. The role assumed by the teacher is that of knowledgeable guide, rather than that of prime dispenser of information and arbiter of what is correct. There is frequent verbal interaction in the classroom as the teacher attempts to stimulate students to produce, explain, and evaluate solutions to problems.

It was not uncommon for a lesson to be organized around a single problem. Asian teachers tended to use short, frequent periods of seatwork, alternating between group discussion of problems and time for children to work through problems on their own. Reflection was a major feature of classrooms.

Brown et al. (1998, p. 373) noted that international observational studies seem to show agreement on some aspects of teacher quality that correlated with students' performance:

- The use of higher-order questions, statements, and tasks that require thought rather than practice
- Emphasis on establishing, through dialogue, meanings and connections between different mathematical ideas and contexts
- Collaborative problem solving in class and small-group settings
- More autonomy for students to develop and discuss their own methods and ideas

In a major study of effective primary school mathematics teaching in Years (grades) 1 to 6 in the United Kingdom, Askew et al. (1997) studied the practices of ninety teachers. A specially designed oral test of numeracy, "tiered" for different age ranges, was administered to the classes of these teachers at the beginning and end of the school year. A total of 2000 children were assessed. Considering relative learning gains, teachers were grouped as "highly effective," "effective," and "moderately effective." Data were then collected (using interviews, questionnaires, and observations) on eighteen "case study" teachers (six in each category), providing information on teachers' beliefs, pedagogical and mathematical subject knowledge, professional development experiences, and practices.

The teaching practices of the highly effective teachers

- connected different ideas of mathematics and different representations of each idea by means of a variety of words, symbols, and diagrams;
- encouraged students to describe their methods and their reasoning, and used these descriptions as a way of developing understanding through establishing and emphasizing connections;
- emphasized the importance of using whatever mental, written, or electronic methods are most efficient for the problem at hand;
- particularly emphasized the development of mental skills.

Possibly just as important were some characteristics of teaching practice that did not enhance students' learning. Teachers who emphasized pupils' acquiring a collection of standard arithmetical methods over establishing understanding and connections produced lower gains. Teachers who gave priority to the use of practical equipment and delayed the introduction of more abstract ideas until they believed a child was ready for them also produced lower gains.

Highly effective teachers had pedagogical and mathematical knowledge and an awareness of conceptual connections among the areas of the primary school mathematics curriculum. There was no particular association between the amount of formal mathematics studied and students' gains. Highly effective teachers were much more likely than other teachers to have undertaken mathematics-specific continuing professional development over an extended period.

The Context of Our Research

The study discussed in this paper took place as part of the Early Numeracy Research Project (ENRP) in Victoria, Australia. Two hundred fifty teachers of grades K–2 participated in a three-year research and professional development project, which explored the most effective approaches to the teaching of mathematics in the first three years of school. The ENRP was a collaborative venture among Australian Catholic University, Monash University, the

Victorian Department of Education and Training, the Catholic Education Office (Melbourne), and the Association of Independent Schools of Victoria. The project was funded in thirty-five project ("trial") schools and thirty-five control ("reference") schools.

There were three important components of this project:

- A research-based framework of "growth points" in young children's mathematical learning (in number, measurement, and geometry)
- A forty-minute, one-on-one interview, used by all teachers with all children at the beginning and end of the school year
- Extensive professional development at central, regional, and school levels, for all teachers, coordinators, and principals

The framework of growth points will not be discussed in detail here (for a fuller description, see Clarke 2001 and Clarke et al. 2003). However, the intention was to draw on available research to describe the typical "learning trajectory" of five- to eight-year-olds.

There were four to six growth points in each mathematical domain. To illustrate the notion of a growth point, we will discuss the domain of *addition and subtraction strategies.* Consider the child who is asked to find the total of two collections of objects. Many young children "count all" to find the total, even when they are aware of the number of objects in each group. Other children realize that by starting at one of the numbers, they can count on to find the total. *Counting all* and *counting on* are therefore two important growth points in children's developing an understanding of addition.

A one-on-one, interactive, hands-on interview was then developed that could furnish classroom teachers with rich information on what their children knew and could do (both individually and as a class) across a variety of domains, with particular insights into the strategies used in solving problems. The disadvantages of paper-and-pencil tests have been well established by Clements (1995) and others. These disadvantages are particularly evident with young children, where reading issues are of great significance. The interview has much to offer the teacher of young children, if time and resources permit.

The task in figure 6.1 enabled the teacher to determine whether a child, given a situation where he or she is required to find out "how many altogether" when shown two groups of known size, chose to count on, chose to count all, used some other successful intuitive strategy, or were unable to solve the problem. (Words in italics are instructions to the interviewer; and normal type shows the words the interviewer uses with the child. "Teddies" refers to small, plastic teddy bears—a major feature of the interview.)

Although the full text of the ENRP interview involved about sixty tasks (often with several subtasks), no child moved through all of them. The interview was in the form of a "choose your own adventure" story, with the inter-

viewer making one of three decisions after each task, as instructed in the interview schedule. After success with a task, the interviewer continued with the next task in the given mathematical domain as far as the child could go with success. If difficulty arose with the task, the interviewer either abandoned that section of the interview and moved on to the next domain or moved into a detour designed to elaborate more clearly the difficulty a child might be having with a particular content area.

At the time of this writing, the interview had been used more than 36,000 times at the K–4 grade levels. For each interview, teachers completed a four-page record sheet. The information on these sheets was then coded by a trained team of coders, assigning achieved growth points to each child for each domain. This process, including statistical measures to convert the growth point data to an interval scale, is discussed in Clarke (2001).

The initial professional development activity was intended to prepare teachers to use the interview. During the remaining three years of the project, the professional development focus was on taking what was learned

18. Counting On

a) Please get four green teddies for me. (*Place nine additional green teddies on the table.*)

b) I have nine green teddies here. (*Show the child the nine teddies, and then cover up the nine teddies with something.*)

That's nine teddies hiding here and four teddies here. (*Point to the groups.*)

c) Tell me how many teddies we have altogether.... Please explain how you worked it out.

d) (*If unsuccessful, remove the cover.*) Please tell me how many there are altogether.

Fig. 6.1. An excerpt from the addition-and-subtraction interview questions

from the interview to help planning and teaching reach their maximum effectiveness, both cognitive and affective.

IDENTIFYING PARTICULARLY EFFECTIVE TEACHERS

Because the project occurred over three years, it was possible to use student interview data from the first two years to identify particularly effective teachers for intensive study in the third year. In identifying effective teachers, we were interested in the *growth* in students' understanding across the school year. Some children, for reasons of home background, language background, or other factors, come to school with less mathematical understanding and skills than others. Our emphasis on growth, however, allowed us to choose teachers who make a difference for all children.

We chose six teachers for case studies. There were two at kindergarten, one at grade 1, one at grade 2, and two teachers of composite grades—a grade K–1 and a grade 1–2 teacher.

One of the kindergarten teachers had made particularly impressive gains in a setting where almost all children were from non-English-speaking backgrounds. The range of grades (K–2) was in recognition that teaching mathematics to kindergarten children is different in several ways from teaching grade 2, for example.

STUDYING WHAT EFFECTIVE TEACHERS DO

The six case-study teachers were studied intensively through the use of the following data sources:

- Five lesson observations by two researchers, incorporating detailed field notes, photographs of lessons, and the collection of artifacts (e.g., worksheets, students' work, teachers' lesson plans)
- Interviews with teachers following the lessons
- Questionnaires completed by teachers through the project
- Teachers' responses to other relevant questions and tasks posed to them

Teachers were observed by two researchers, working together for five lessons. Lessons on three consecutive days were observed in the middle of the school year, and then lessons on two consecutive days were observed a couple of months later. Teachers were asked to focus on different content, of their choosing, in each of the two sets of lessons (e.g., number in the first set, geometry in the second). Both observers used laptop computers to take notes on the lesson, and teachers were interviewed after each lesson (interviews were audiotaped and transcribed), to discuss their intentions for the lesson and what had transpired. Many photographs were also taken by the researchers to give a richer picture of the classroom environment and the activities used. In all, seven observers made a total of eighty-six visits to schools during the case studies, which included time for intensive practice of the proposed techniques for data collection. The children and teachers had become used to being visited by members of the research team over the course of the project, and so they became comfortable with the observers quite quickly and continued their normal routines.

Decisions needed to be made on the kinds of notes that would be taken during lessons. It was decided to attempt to note as much as possible of what transpired in the lesson in a relatively "free" form. We were guided in this decision by the experience of others. For example, Stigler and Baranes (1988) conducted mathematics classroom observations in three countries using two methods. In the first, a structured coding scheme was used with an elaborate time-sampling plan. It was therefore possible to obtain estimates of the amount of time allotted to different classroom activities. In the second study, the researchers "decided to trade the greater reliability of an objective coding scheme, for the inherent richness of detailed narrative descriptions of mathematics lessons" (p. 294). The ENRP research team made the same decision.

We did, however, agree on a framework for the observations and interviews (see fig. 6.2). This framework was chosen to be quite broad, anticipating a criticism that the research team might have constrained their observations because of what they *hoped* they would see. We believe that most people with an interest in mathematics education, no matter what their philosophical persuasion, would agree that these categories were reasonable components to study. Our aim was to describe the practice of demonstrably effective teachers and ultimately to look for common themes, not to judge.

At several stages during the case study, the team met to describe to one another what they were seeing. "Critical friends," not involved in the research, gave feedback on the kinds of themes they were hearing as verbal reports were made. The first three lessons and the subsequent discussion prompted the team to focus on particular aspects in the last two lessons and interviews that had not necessarily been noted to that point but that had emerged in the team's discussions.

Mathematical focus
Features of tasks
Materials, tools, and representations
Adaptations, connections, links
Organizational style(s) and teaching approaches
Learning community and classroom interaction
Expectations
Reflection
Assessment methods

Fig. 6.2. Categories within the ENRP lesson observation and analysis guide

WHAT CAN BE SAID ABOUT EFFECTIVE TEACHERS?

Following the lesson observations and interviews and a number of the research team's meetings, it was decided to use the original framework to describe the practices of effective teachers. It was agreed to list common elements where evidence was available for *at least four of the six teachers.* The description of effective teachers, as revealed in this study, is given in figure 6.3. It will be noted that an additional category, "personal attributes," was added at this stage, since these attributes had emerged strongly from the data.

The twenty-five common elements of good teaching sit well with the Teaching Principle (NCTM 2000): "Effective mathematics teaching requires understanding what students know and need to learn and then challenging and supporting them to learn it well" (p. 16).

WHAT DOES THIS ADD TO OUR KNOWLEDGE ABOUT EFFECTIVE TEACHERS?

As the reader considers this list (fig. 6.3), several points can be made:

- The six teachers were chosen from 250 on the basis of their students' growth in understanding and skills over two years of teaching, using an instrument in which we can have great confidence. The one-on-one interview provided far richer and more important data than can possibly be revealed by standardized testing, particularly in grades K–2. Therefore, we can be confident that the choice of teachers was made according to the things we value.
- Although only six teachers were studied intensively, our visits to the classrooms of other project teachers (578 school visits in all) led us

Effective teachers of K–2 mathematics ...

Mathematical focus

- focus on important mathematical ideas
- make the mathematical focus clear to the children

Features of tasks

- structure purposeful tasks that enable different possibilities, strategies, and products to emerge
- choose tasks that engage children and maintain their involvement

Materials, tools, and representations

- use a range of materials, representations, and contexts for the same concept

Adaptations, connections, links

- use teachable moments as they occur
- make connections to mathematical ideas from previous lessons or experiences

Organizational style(s) and teaching approaches

- engage and focus children's mathematical thinking through an introductory, whole-group activity
- choose from a variety of individual and group structures and teacher roles within the major part of the lesson

Learning community and classroom interaction

- use a range of question types to probe and challenge children's thinking and reasoning
- refrain from telling children everything
- encourage children to explain their mathematical thinking and ideas
- encourage children to listen to and evaluate others' mathematical thinking and ideas
- listen attentively to individual children
- build on children's mathematical ideas and strategies

Expectations

- have high but realistic mathematical expectations for all children
- promote and value effort, persistence, and concentration

Reflection

- draw out important mathematical ideas during or toward the end of the lesson
- after the lesson, reflect on children's responses and learning, lesson activities, and lesson content

Continued

Assessment methods

- collect data by observation or listening to children, taking notes as appropriate
- use a variety of assessment methods
- modify planning as a result of assessment

Personal attributes of the teacher

- believe that mathematics learning can and should be enjoyable
- are confident in their own knowledge of mathematics at the level they are teaching
- show pride and pleasure in individuals' success

Fig. 6.3. Common themes emerging from the ENRP case studies of effective teachers

to believe that the features in figure 6.3 were increasingly evident over the three years of the project in other classrooms as well. Those teachers took what they had learned from the interviews about children's mathematical thinking and, working with colleagues, endeavored to provide the kinds of activities and tasks that enhanced learning for all students. In a questionnaire at the end of the project, all teachers were asked to identify the greatest changes in their teaching practice. Among the most common were the use of more open-ended tasks and activities; more probing questioning, asking why and how, and valuing children's thinking; challenging and extending children's thinking and having higher expectations; more practical, hands-on activities; and greater emphasis on reflection and sharing.

- The fact that a study in Australia yielded findings that sit so comfortably with recommended classroom practices articulated in the NCTM's *Principles and Standards for School Mathematics* is encouraging for teachers in both Australia and North America.

SNAPSHOTS FROM THE CLASSROOMS OF EFFECTIVE K–2 TEACHERS

Sometimes a list such as that in figure 6.3 can seem a bit removed from the life and color of the classroom. In order to give a picture of the vibrant kinds of learning communities we observed during the project, we now discuss five themes from the list.

Effective teachers structure purposeful tasks that enable different possibilities, strategies, and products to emerge

The interviews made very clear the considerable range of knowledge and understanding within any classroom. As a result, teachers made extensive use of *open tasks*, encouraging children to share their strategies of solution. Examples of such tasks include these:

- I rolled three dice and the total was 10. What might the dice have been?
- Two children measured the basketball court with rulers. Huong said it was 20 rulers long. Yasmine said it was 19. Why might that be?
- I drew a shape with four sides. What might my shape look like?
- I counted something in the room. There were exactly four. What might it have been?
- I bought some things at the supermarket and got 35 cents in change. What did I buy and how much did each item cost?
- What can you find that is lighter than a potato but bigger than it?

Children eagerly accepted the challenge of these kinds of tasks, responding at their own level of understanding.

Effective teachers of mathematics focus on important mathematical ideas and make the mathematical focus clear to the children

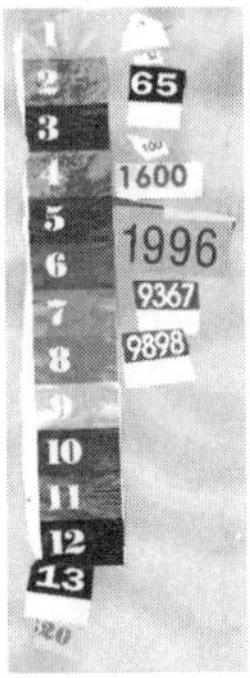

Teachers noticed during the interview that although many children could read and write two- and three-digit numbers, quite a few had difficulty ordering numbers, even one-digit ones. In school teams, they developed a range of games and activities that focused on this important mathematical idea. One teacher asked children to cut up magazines and catalogs, taking out any numbers they could find. They then sorted these from smallest to largest. The example shown in the photograph illustrates the potential range of responses from children, from small numbers to quite large ones, due to the openness and accessibility of the task.

Another teacher used a card game, where the picture cards were removed from a standard deck, and two children each had half the deck. Each child turns over a card at the same time, and the person with the larger of the two numbers takes both. Once again, there was a clear focus on ordering numbers.

Effective teachers choose tasks that engage children and maintain their involvement

The children in a kindergarten class were "celebrating" 100 days at school, with a range of activities that focused on 100. They were shown how to make paper chains from strips of paper. They were then taken outside, and each table group had to draw a chalk line that they believed would be a good estimate of how far 100 such "links" would stretch, measured from the classroom window.

Each group then had the task of creating such a chain. Of course, keeping track of how many links they had made and the fine motor skills involved in the assembly posed a challenge for these children. But all were motivated and involved, and after one hour they were ready to check how closely the chains matched their estimates.

This lesson was evidence of a teacher who had high but realistic expectations of her pupils, and it also promoted and valued effort, persistence, and concentration. In using the same activity at the grade 1 level, another teacher encouraged the children to use the colors of the links in some way to make it easier to count the links. One group decided to use patterns of five links for this task.

Effective teachers of mathematics encourage children to explain their mathematical thinking and ideas and build on children's mathematical ideas and strategies

One of the benefits of regular use of the interview was that the kinds of tasks used and questions posed furnished models of the kinds of probes that could be used in classrooms. Teachers commented that they found themselves using many more questions that probed children's thinking. Examples included the following:

- How did you work that out?
- Could you do that another way?
- How are these two objects the same, and how are they different?
- What happens if I change this here?
- What could you do next?
- Can you see a pattern in what you've found?
- Can you make up a new task using the same materials?

Effective teachers use teachable moments as they occur

A teacher who taught a mixed kindergarten and grade 1 class used children's literature regularly in her mathematics classes. One morning, she read *Counting on Frank* (Clement 1994) to the children.

After she read the last page, she asked the children, "Who can remember how many jelly beans were in the jar?" One of the children said that it was seven hundred and something, and so the teacher said, "I am going to read it again, and I want someone to write it on the board." A child wrote 70045.

The teacher commented that the child had done a good job but noted that although this was how we *read* it, we *write* it a different way. The next child

wrote 7045, and a third wrote 745. The teacher had some Montessori cards beside her chair, and she showed the class the units and the 10s. Next, she showed the 100s and had the children read some of them and talk about the number of digits. She asked one of the children to make 745 with the cards while she showed the rest of the class 1000s cards. She made several numbers, such as 7020, asking the children to read the numbers as she made them.

She asked the children if they had seen numbers like these before, and after mostly blank looks, she pointed to the board. "What year is it?" she asked. 2001 was written as part of the date on the board. This teacher took the opportunity of a child's understandable but incorrect response and explored mathematical ideas that were beyond the usual expectations for five- and six-year-olds but clearly linked back to their experiences.

She was not expecting full understanding at that point but was exploring within the children's experience mathematical concepts that would be revisited in the future. Using materials that appropriately focused on overcoming the misconception also illustrates a teacher who really understands the mathematics that these children need to develop and who has a range of materials at her disposal to enable an appropriate choice as the moment arises. An opportunity was taken to move in a mathematically purposeful direction.

CONCLUSION

It is interesting to consider the extent to which the twenty-five teacher behaviors and characteristics in figure 6.3 have application to other grade bands. We believe that similar research in grades 3–12 (and possibly beyond) would yield many elements in common with these. Indeed, a discussion of this topic would be most worthwhile among preservice and in-service teachers.

It was a privilege to be in the classrooms of dedicated mathematics education professionals. We have attempted to describe their practice in ways that will "ring bells" for readers. We have described activities and teaching strategies where the enthusiasm, curiosity, and strategies of young children are valued and developed, with lasting effects on their understanding, their attitudes, their love of mathematics, and their confident views of themselves as learners of mathematics.

REFERENCES

Askew, Mike, Margaret Brown, Valerie Rhodes, Dylan Wiliam, and David Johnson. "Effective Teachers of Numeracy in UK Primary Schools: Teachers' Beliefs, Practices and Pupils' Learning." In *Proceedings of the Twenty-first Conference of the International Group for the Psychology of Mathematics Education*, Vol. 2, edited by Erkki Pehkohen, pp. 25–32. Helsinki, Finland: Lahti Research and Training Centre, University of Helsinki, 1997.

Brophy, Jere E., and Thomas L. Good. "Teacher Behavior and Student Achievement." In *Handbook of Research on Teaching*, edited by Merlin C. Wittrock, pp. 328–75. New York: Macmillan, 1986.

Brown, Margaret, Mike Askew, Dave Baker, Hazel Denvir, and Alison Millett. "Is the National Numeracy Strategy Research-Based?" *British Journal of Educational Studies* 46 (December 1998): 362–85.

Clarke, Doug M. "Understanding, Assessing and Developing Young Children's Mathematical Thinking: Research as a Powerful Tool for Professional Growth." In *Numeracy and Beyond: Proceedings of the Twenty-fourth Annual Conference of the Mathematics Education Research Group of Australasia*, edited by Janette Bobis, Bob Perry, and Michael Mitchelmore, Vol. 1, pp. 9–26. Sydney, New South Wales, Australia: Mathematics Education Research Group of Australasia, 2001.

Clarke, Doug M., Jill Cheeseman, Andrea McDonough, and Barbara Clarke. "Assessing and Developing Measurement with Young Children." In *Learning and Teaching Measurement*, 2003 Yearbook of the National Council of Teachers of Mathematics (NCTM), edited by Douglas H. Clements, pp. 68–80. Reston, Va.: NCTM, 2003.

Clement, Rod. *Counting on Frank.* Sydney, New South Wales, Australia: Angus & Robertson, 1994.

Clements, Ken A. "Assessing the Effectiveness of Pencil-and-Paper Tests for School Mathematics." In *Galtha: Proceedings of the Eighteenth Conference of the Mathematics Education Research Group of Australasia*, edited by Bill Atweh and Steve Flavel, pp. 184–88. Darwin, Northern Territory, Australia: University of the Northern Territory, 1995.

National Council of Teachers of Mathematics (NCTM). *Principles and Standards for School Mathematics.* Reston, Va.: NCTM, 2000.

Stigler, James W., and Ruth Baranes. "Culture and Mathematics Learning." In *Review of Research in Education,* edited by Ernst Z. Rothkopf, pp. 253–306. Washington, D.C.: American Educational Research Association, 1988.

Stigler, James W., and Harold W. Stevenson. "How Asian Teachers Polish Each Lesson to Perfection." *American Educator* 15 (spring 1991): 12, 14–20, 43–47.

7

The NCTM *Standards* in a Japanese Primary School Classroom

Valuing Students' Diverse Ideas and Learning Paths

Aki Murata

Naoyuki Otani

Nobuaki Hattori

Karen C. Fuson

CREATING a learning environment where individual students' ideas, interests, and experiences are valued is important. It can help inform teachers where their students are and how to guide the class so that their students develop a good understanding of the content while expressing their own thinking. However, creating such an environment is not a simple matter. In the classroom context, where teachers are expected to meet the needs of students of diverse backgrounds while adapting themselves to the changing contexts of teaching, the complexity of the challenge can be overwhelming.

The result of the Third International Mathematics and Science Study (TIMSS) indicated that Japanese elementary school mathematics classrooms reflected many characteristics of U.S. reforms designed to develop such learning environments. The open-ended problem solving in Japanese classrooms and discussion-centered approaches effectively draw students' ideas out into the open where teachers can facilitate students' understanding (TIMSS 1997). Several comparative studies have also demonstrated that

The research reported in this article was supported in part by the National Science Foundation (NSF) under grant no. REC-9806020 and in part by the Spencer Foundation. The opinions expressed in this paper are those of the authors and do not necessarily reflect the views of NSF or the Spencer Foundation.

Japanese classrooms are qualitatively different from U.S. classrooms in both teachers' expectations of what learning is and students' academic performance (e.g., Inagaki, Morita, and Hatano 1999; Stigler and Hiebert 1999). Japanese teachers are guided by the beliefs that all students will be able to do mathematics with enough time and practice and that individual developmental differences must be recognized. Consequently, their questions are sequenced to become gradually more complex as students gain experiences that support their learning. A safe classroom environment helps students share their ideas, make mistakes, and learn together. This qualitatively different classroom image closely reflects many ideas presented in the National Council of Teachers of Mathematics (NCTM) *Principles and Standards for School Mathematics* (NCTM 2000).

In many U.S. classrooms where teachers are serious about incorporating reform ideas in their teaching, we see images of similarly effective teaching practices (e.g., Fuson et al. 2000; Lampert and Ball 1998). This article extends our existing knowledge of effective teaching practice by examining aspects of Japanese teaching that are closely aligned with the Teaching Principle (NCTM 2000). Seeing the U.S. reform ideas in action in a Japanese classroom enables U.S. teachers to see how the elements of the Teaching Principle are practiced in a different cultural context, how Japanese teachers help students learn, and how certain aspects of Japanese practice might be incorporated in instruction in the United States.

We discuss (*a*) how valuing students' diverse ideas and learning paths helps create an effective mathematics-learning environment and (*b*) the roles of different visual learning supports in such an environment. Although we are using an example from grade 1, the ideas are applicable throughout the elementary school grades.

SCHOOL AND PARTICIPANTS

Green Leaf Community School is a full-day Japanese school in a suburb of Chicago. It is one of the two full-day schools in the United States that are operated by the Japanese Ministry of Education and that closely follow the Japanese National Course of Study. Administrators and teachers are sent directly from Japan through the ministry, and the instructional language is Japanese. The second and third authors of this article are Japanese teachers who taught at the school. The school houses approximately 200 students in first through ninth grades. These students typically come to the United States because of their fathers' work in the area, and the families who relocate under these circumstances typically stay in the United States for two to five years and then return to Japan. They are different from other ethnic minority groups in the United States who immigrate to stay. This Japanese community puts much effort into preserving the Japanese culture in their children's lives because they wish for their children to be successful on returning to Japan. For that reason, maintaining Japanese ways of teaching and learning is considered to be important.

This paper summarizes the intensive data collected in one grade 1 classroom of twenty-five students as they engaged in learning teens additions. This is one of the core units of grade 1, and it was typical of the teaching observed throughout the year and in other grade levels. Learning the break-apart-to-make-ten method is viewed as crucial for later multidigit calculation because it enables students to think about teen numbers as "10 + another number" to make regrouping ("carrying") more sensible. The unit was taught during the fifth month of the school year and consisted of eleven lessons over a three-week period. Since the beginning of the school year, grade 1 students had explored numbers and counting up to 10, decomposition and recomposition of numbers smaller than 10, addition and subtraction of numbers with 10 to make teens without regrouping (e.g., $10 + 3 = 13$, $18 - 8 = 10$), and addition and subtraction with three numbers (e.g., $8 + 2 + 4 = 14$, $17 - 7 - 2 = 8$).

Learning to group numbers in tens may be considered important mathematically across cultures because of the worldwide base-ten system. However, in Japan, making tens is also culturally important, and Japanese students have opportunities outside the classroom to experience the "ten-ness" in their everyday lives. The Japanese language, like other Chinese-based languages, explicitly states "ten" in teen numbers (e.g., 11, 12, 13 are said as "ten-one," "ten-two," and "ten-three") and in the decade numbers (e.g., 50 is said as "five ten"), and thus students hear and experience ten within numbers. Japanese students may feel the "bumps" as they hear the numbers "ten-one," "ten-two," and "ten-three," and this gives them a clue that after ten, a new sequence starts as the counting in the ones column starts over (Easley 1983; Fuson and Kwon 1992).

Students also see things typically grouped in tens to be sold in stores in Japan, such as eggs in a carton of ten (instead of a dozen) and bottled drinks are packaged as a ten-pack instead of a six-pack. Smaller packages of five items are also available, such as postcards, and pairs of socks may be sold in packages of five. Because of their use of the metric system, too, students hear and experience how groupings of ten are used for measuring their everyday items. Such experiences contribute to young Japanese students' development of a ten-structured understanding of quantities. The learning of the break-apart-to-make-ten method in grade 1 is considered an extension of students' experiences of ten-ness in their everyday lives—experiences that support the development of their future place-value understanding. Japanese students' experiences are more uniformly consistent than those of their U.S. peers, who use the base-ten system in classrooms but who have their everyday experiences with quantities that depend on different reference numbers (e.g., 12 for dozen, 16 for weight conversion between pounds and ounces).

Although the classroom was observed during the teens addition unit, careful field notes were taken to record (1) the general lesson procedures, (2) teaching and learning activities and their structures, (3) the kinds of student participation in the activities, (4) students' engagement and reaction to the activities, (5) verbatim transcripts of whole-class and small-group discussions (as many as possible of the latter), (6) questions and responses of teacher and students, and (7) conversations among researcher, teacher, and students before, during, and after the class. The lessons were also videotaped to support and supplement the observation notes. Classroom artifacts (copies of worksheets, tests, quizzes, etc.) were also collected.

VALUING STUDENTS' IDEAS

As the lessons unfolded in the teens addition unit, the teacher carefully orchestrated discussions so that students shared their different ideas that were based on their prior understanding of the decomposition of numbers (e.g., 4 is decomposed into 1 and 3). The major parts of the first two lessons were devoted to students' sharing various methods to understand the addition situation of "9 + 4" by decomposing two addends into different quantities that were easier to add. The unit followed the National Course of Study and was designed to introduce grade 1 students to "9 + another number" problems first and then proceed to "8 + another number," then to "7 + another number" problems. This helped students to extend their prior understanding of the decomposition of 10 by focusing on certain break-apart pairs (e.g., 9 and 1, 8 and 2). Instead of focusing on the same total number with different addends to make that total (e.g., 13 for 9 + 4, 8 + 5, 7 + 6), the Japanese approach of grouping problems with the first addend helps students do the first step of the break-apart-to-make-ten method.

For the initial introduction, the teacher showed a group of nine blue magnetic counters and a group of four red magnetic counters on the chalkboard (fig. 7.1). Some students immediately shouted out the answer, "13!" The teacher then initiated discussion by saying, "Some of you are quick in telling the answer, but who can share with the class your thinking?"

Several students raised their hands to share their ideas. Sakiko went up to the board when called on and moved one red counter to add to the blue group.

Sakiko: From 4, I add 1 to 9 and make 10.

Teacher: So, the 9 became 10 and the 4 became 3?

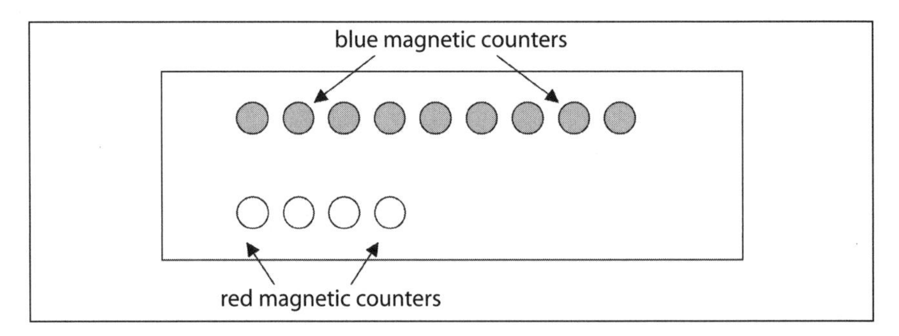

Fig. 7.1. Magnetic counters on the board to guide students' thinking for 9+4

Sakiko: Then we know 10 and 3 make 13. We learned that before [*in a previous unit*].

Tadashi then volunteered to show the class how he counted unitarily, 1 to 13, to get the answer. Following his contribution, several students shared different counting methods: Tetsuhiro counted by twos, and Tomoko counted by threes. Although counting was not the official topic of the lesson, the teacher allowed students time to explain, and he then summarized to the whole class how counting by ones, twos, and threes were different from one another.

Nobuhiko raised his hand. He came to the board to show how he solved the problem, 9 + 4.

Nobuhiko: I made 9 into 7 and 2, and then thought 7 and 6 is 13.

Teacher: You always think differently. [*Friendly laughter from the class*] Thinking 7 and 6 is easier for you?

Nobuhiko: Yes!

Nobuhiko then took the two counters on the right in the top row and moved them down to align them with the counters on the bottom. He then counted them one by one to make sure there were 7 counters on the top and 6 on the bottom.

The teacher then called on Koichi. Koichi is one of the fastest students in the classroom and always seems to have good ideas. He came up to the board, separated counters, and made three groups of 5, 5, and 3.

Koichi: I thought this way: 5 and 5 is 10; so, 3 more is 13.

The teacher then summarized the different approaches shared by students. The making-ten method was called the "Sakiko method," changing 9 + 4 into 7 + 6 was called the "Nobuhiko method," and changing 9 + 4 into 5 + 5 + 3 was called the "Koichi method." The different counting methods were named for the students who shared those particular methods. The Sakiko method (break-apart-to-make-ten method) was the primary method to be taught in this unit according to the curriculum, but the teacher spent approximately equal amounts of time explaining each method at this point by asking students questions.

The teacher then extended his students' exploration of methods by using the new addition situation of "8 + 5" (this requires students to start with a different break-apart pair for 10: 8 and 2). He asked students to apply the methods they had come up with as a class to solve the new problem.

> *Teacher:* Who can explain again how to solve this using the Sakiko method?
>
> *Sanshiro:* We can separate 5 into 2 and 3, and make 10 with 8 and 2, then add 3 to make 13.

The teacher summarized Sanshiro's thinking on the board while verbally reviewing the steps by asking students. After repeating the same process with the other student methods, the teacher asked the students to vote for the method they thought was the easiest to use in this addition situation. Sakiko's method of making-ten received the most votes, and all other methods also got several votes.

In this first lesson of the unit, the teacher welcomed students' different ideas and allowed time for diverse methods. The sharing process was important at the beginning of the unit because it gave an opportunity for students to review previously learned concepts, demonstrate their competency, and set the stage for future exploration. When some students immediately stated the answer for the problem, the teacher recognized their experience and knowledge, yet redirected the students' focus to the process of solving the problem, thus offering opportunities for the students who were experiencing addition with totals in teens for the first time. He then stated the importance of clear explanations to show their thinking with the visual representational supports on the board (see the following section for the discussion of the visual representational supports). When several students shared more basic counting methods, the teacher recognized and valued their contributions even though the focus of the lesson was not officially on counting.

The students' sense of ownership in learning was created and maintained throughout the lesson. The teacher valued the diverse thinking of students by asking them to share their different methods, maintained their ownership of ideas by referring to the methods using the students' names, and encouraged critical thinking by having different students explain and apply the method in different addition situations. Although voting in this initial lesson divided students into different camps, this voting for "the easiest method" continued on to subsequent lessons until all students agreed that "making of ten" was the most useful method. Voting helped individual students express their choices, and with the teacher's gentle guidance, students gradually made their own decisions for their learning.

VISUAL LEARNING SUPPORTS FOR EACH STEP IN THE BREAK-APART-TO-MAKE-TEN METHOD

The teacher used visual representational supports to emphasize the different decomposition steps students had taken in their methods. For the target break-apart-to-make-ten method, he initially supported each step visually. As the unit progressed, he gave support for fewer steps, and at the end only the most difficult step was shown visually. The whole process of the break-apart-to-make-ten method is summarized in figure 7.2.

When the teacher first introduced the visual representational support, he used the extensive Level A questioning (fig. 7.2) to guide students carefully through all four necessary steps while using large counters on the board to assist thinking (fig. 7.1). He effectively combined his questions, the counters on the board, the drawn visual representational support with numbers, and gestures to help students make connections among the different representations and different steps.

The teacher also explained to students that the reason they used the visual representations was to show their thinking in the open. He emphasized that it is important for students to explain their thinking clearly even though he said, "Only answers would be easier." At the beginning of the unit, to draw students' attention to the ten-structured way of seeing the number, the teacher carefully used red chalk to encircle the two numbers that made 10 (steps 3 and 4 of fig. 7.2). Students were also encouraged to use red pencils to circle numbers during the first week in their notebooks. After the first week, the teacher changed to using only white chalk to draw the whole thing. Using the visual representational support to solve problems in their notebooks was very helpful, since it enabled students to keep track of each step of the procedure. Students gradually came to experience the break-apart-to-make-ten method as a fluent whole. From the second week, the teacher allowed each individual student to decide whether he or she would draw the visual representational support or not in his or her notebook. However, whenever he saw students having a problem, he encouraged them to draw the visual representational support to check the thinking procedures.

Students also used smaller versions of the counters the teacher used on the board. For the first three lessons, containers of counters were distributed to each student at the beginning of the class, and students used them to explain their thinking to peers during independent seatwork time. From the fourth class, the teacher kept the containers of counters with him and told students that counters would be available to them if needed. As students became familiar with the visual representational drawings, fewer of them chose to use counters. However, when the teacher saw some students taking a long time or struggling with a certain problem, he suggested that they use the counters to review the process and guide their thinking.

Steps / Levels	1. Realize that 9 needs 1 more to make 10	2. Separate 4 into 1 and 3	3. Add 9 and 1 to make 10	4. Add 10 and 3 to make 13
Level A	"9 and what number make 10?" (Teacher points to 9.)	"What two numbers are you separating 4 into (to make 10)?" (Teacher draws to show break-apart pairs for 4.)	"What do 9 and 1 make?" (Teacher circles 9 and 1 and writes 10 next to the circle.)	"What do 10 and 3 make?" (Teacher points to numbers 10 and 3 and says, "Ten and three" to make connections to "ten-three."
Level B		"What two numbers are you separating 4 into (to make 10)?" (Teacher draws to show break-apart pairs for 4.)	"9 and 1 make …?" (Teacher points to 9 and 1.)	"10 and 3 make …?" (Teacher points to 3.)
Level C		"4 is what number and what number?" (Teacher draws to show break-apart pairs for 4.)		"10 and 3 make …?" (Teacher points to 3.)
Level D		No verbal guiding, teacher uses visual support of drawing break-apart pairs for 4.		

Level E (No guiding of steps)				
Visual representational support in textbooks and on the board	**9** + 4 / 1	9 + **4** / 1 3	(**9**)+ 4 / 10 **1** 3	(**9**)+ 4 / **10** 1 **3**

Note: The increasing levels support fewer steps.

Japanese number words, as in other Chinese-based languages, support making ten. They explicitly state "ten" in teen numbers (e.g., "eleven," "twelve," and "thirteen" are said "ten-one," "ten-two," and "ten-three"). Japanese children experience 10 within teen numbers as they hear the "bumps" in the numbers such as "ten-one," "ten-two," and "ten-three." This gives them a clue that after 10, a new sequence starts as the counting in the ones column starts over.

Fig. 7.2. Levels of teacher support during the teens additions unit for learning the break-apart-to-make-ten method

Students also used their fingers as counters. The teacher never discouraged the use of fingers and used them himself sometimes to show students the break-apart pairs of certain numbers. As with counters, students used their fingers less frequently as they came to use the visual representational drawings. The teacher suggested the use of fingers to slower students to confirm break-apart pairs of certain numbers or the amount needed to make ten.

VALUING STUDENTS' DIFFERENT LEARNING PROCESSES

As the unit progressed, students had many opportunities to practice the making-ten process in the classroom: as a whole class orally, as individuals orally in a whole-class situation, as pairs doing independent seatwork, and as individuals doing independent seatwork. Every lesson started with a whole-class review of the break-apart-to-make-ten method, and the teacher's questions guided the students to support their growing competency. The new addition steps were not difficult when they were taken one step at a time, and students developed their understanding of, and fluency with, the overall process as the unit progressed.

All students had opportunities to solve problems individually in the whole-class context supported by questions from the teacher. As individual students gradually came to use the break-apart-to-make-ten method, the different levels of understanding became apparent in the whole-class context. Although the teacher progressed during the unit to supporting fewer steps, when students hesitated or gave wrong answers, he would fall back from a higher level to Levels A, B, or C (see fig. 7.2), supporting each step at that level with questions. Sometimes this "retreat" was used for the whole class, and sometimes it was for individual students.

For example, in the fifth lesson of the unit, the teacher wrote a set of questions on the board that included "9 + another addend" and "8 + another addend" addition problems (fig. 7.3). His initial expectation was that students would state the answers to those questions quickly (Level E support). However, when he saw the first student hesitate, he quickly realized that the students might not be ready for this rapid exercise and changed his expectations. He asked the students to do the hardest step for every problem, finding the needed break-aparts, and he wrote break-apart pairs on the board (Step 2, fig. 7.2).

> *Teacher:* Is this hard? OK, let's do it this way. For 9 + 2, we will separate 2 into what two numbers?
>
> *Yutaro:* 1 and 1.

When all the break-apart pairs had been written on the board under the second addends, the teacher asked the whole class to say the answer to each question together orally as he pointed to it with a long ruler (Level C support). With the visual break-apart support on the board, students were more confident as they stated the answers together quickly. The teacher then asked individual students to repeat the same procedure in a whole-class context. Most were able to carry out the final two steps on their own, but there were several students who needed additional support. For example, as one student hesitated on her turn, the teacher quickly changed his approach.

> *Teacher:* [*Points to 8 + 4*]
>
> *Masayo:* 8 + 4 is ... 18? 19?
>
> *Teacher:* Let's slow down and take time here, Masayo [*signaling that it is important to think and that thinking takes time*]. We have 8 + 4. How do we separate 4? (Supporting Step 2)
>
> *Masayo:* 2 and 2?
>
> *Teacher:* Then, we have 8 and 2? (Supporting Step 3)
>
> *Masayo:* 10, so 2 more is 12 (did Step 4 independently).

The whole-class practice context was very important when students shouted out the answers to the teacher's questions together or individually. It offered opportunities for students to share their growing expertise together with their classmates, and it enabled the teacher to assess individual students' progress and assist those who needed it. It was also an enjoyable expe-

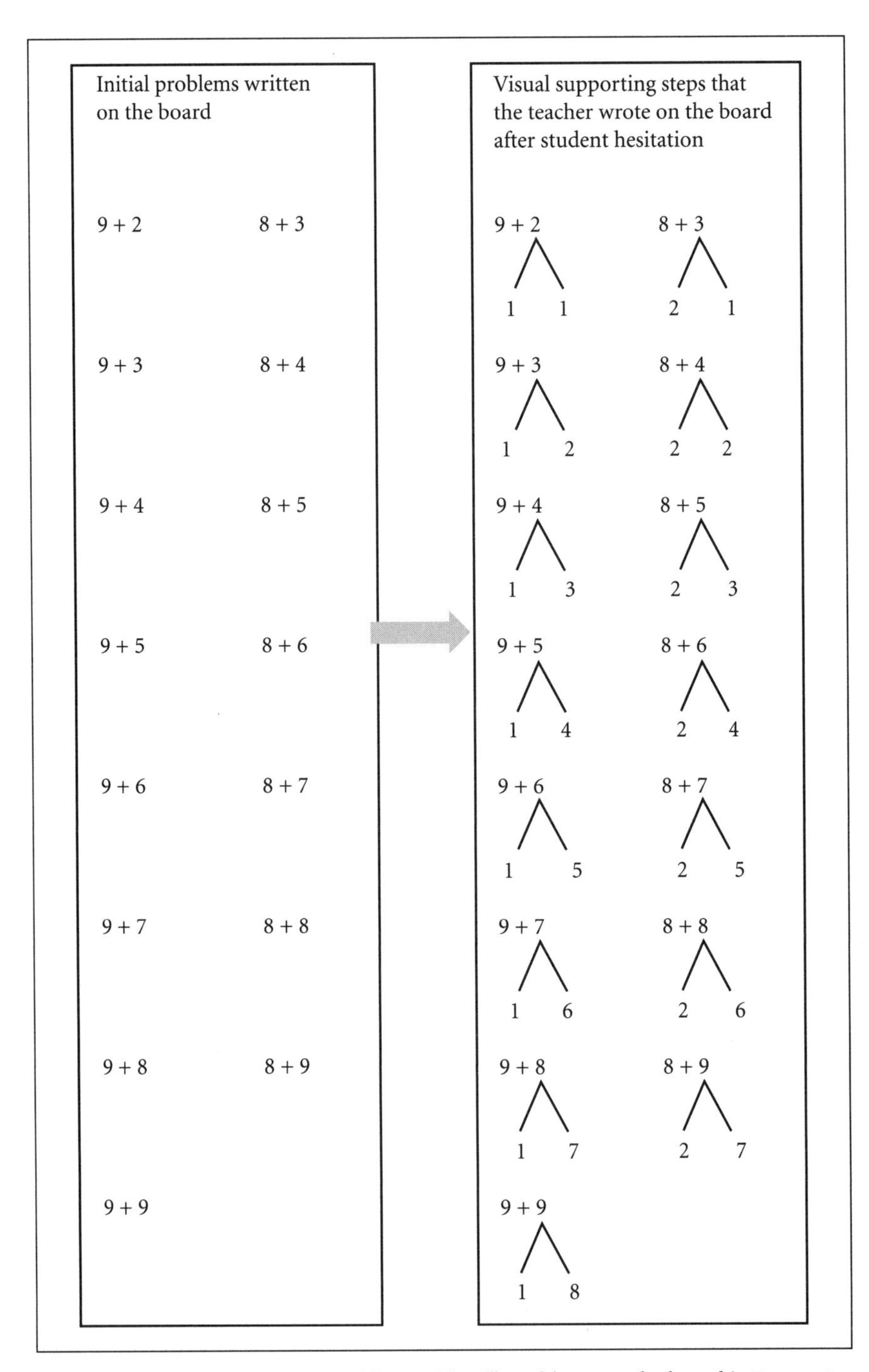

Fig. 7.3. Using "9 + addend" and "8 + addend" problems on the board in Lesson 5

rience where everyone supported one another. Each time an individual student gave an answer to the teacher's question, the student typically asked the whole class, "Is it OK?" When the answer was acceptable, the whole class responded, "It is OK!" to recognize the student's contribution. When the answer was not acceptable, the whole class responded, "It is not OK!" Such a response was typically followed by additional supporting questions from the teacher to trace the steps that the student had taken to arrive at that incorrect answer. This classroom ritual placed students in the role of evaluating one another's mathematical contribution, and the teacher worked to support the evaluation. The feeling of "togetherness" motivated individual students to move forward in the classroom community, and because the slower students were supported as necessary by the teacher, everyone could experience the whole addition process together repeatedly.

This feeling of togetherness was created gradually from the beginning of the school year. Students were given plenty of opportunities to play and work together during the school day both in and outside the classroom. The teacher often supervised student activities indirectly so that students would be responsible for making their own decisions. He also sometimes intentionally created situations where students needed to find ways to work together and solve problems themselves. The teacher believed that his main job was to help students find solutions to their problems, not to give answers. This belief was reflected not only in the way he guided students' learning through questioning but also in the way he supported and guided students' own negotiations in resolving playground conflict. Small-group work was common in the classroom, and cooperation was expected in every aspect of their school lives. Through many successful and not-so-successful experiences of trying to work together, students gradually developed an understanding of one another. They became aware of the strengths and weaknesses of their peers and developed ways to work to support one another. This interdependence among members of the classroom community generated a sense of belonging, and students worked hard to maintain this feeling of togetherness in the community.

Every student was expected to do math well in the classroom. Although the teacher obviously recognized individual developmental differences and adjusted his questions to support the differences, he created a classroom environment that offered ample opportunity for students to experience the concepts and practice the break-apart-to-make-ten addition method together. The existence of different learning paths in the classroom was not viewed as a problem but as a way to help students at different levels. The slower students were involved in the rapid whole-class oral practice through the support of the teacher and thus experienced the whole scope of the process along with the faster, fluent students; the faster students were given time and opportunities to review each step of the method with the slower students.

CONCLUDING REMARKS

The example of the Japanese first-grade classroom presented in this article shows (1) how valuing diverse ideas and learning paths worked to create an effective learning environment in the classroom and (2) how different learning supports helped students learn. Students did not learn the concepts by rote. Instead, in the sense-making environment that supported students' ideas, they remained active agents in their own learning as they followed their own learning paths. The aspects of effective teaching for understanding observed in this situation were also observed in other grades in the school. We summarize them in figure 7.4.

Valuing students' ideas and thinking	• Encourage students to share ideas and different approaches. • Ask students questions to guide their thinking. • Maintain students' ownership of ideas (name different methods with students' names, vote for different methods but discuss advantages and disadvantages).
Providing visual representational support for learning steps	• Use representations (physical objects, drawings, fingers, oral explanations) to strengthen students' understanding. • Make connections among different representations. • Guide students' thinking with visual cues (e.g., circling numbers to make 10, drawing upside-down V to show break-apart pairs) to suggest crucial steps in the process. • Emphasize the critical step of the procedure by using colored conceptual drawing.
Valuing and supporting students' different learning paths	• Vary questioning patterns to meet the different levels of understanding of individual students. • Include slower students in whole-class practice to allow them to experience the entire process rapidly but support them as necessary with questions. • Consider differences among students as strengths, and create situations where they benefit from the differences.

Fig. 7.4. Aspects of teaching for understanding in the Japanese classroom

Teaching is a complex task that helps shape students' perspectives toward mathematics for their future, and students develop their understanding through the experiences provided by their teachers in the classroom. Seeing an example of an effective learning environment that supports students' diverse ideas and learning paths in a Japanese classroom may help us understand how the aspects of teaching that are presented in the NCTM *Standards* work together to create an environment that supports students' engagement, sharing of ideas, and learning. Teachers can help all students learn both mathematically and culturally valuable methods in a meaningful and sensitive way adapted to students' individual learning paths.

REFERENCES

Easley, John. "A Japanese Approach to Arithmetic." *For the Learning of Mathematics 3,* no. 3 (1983): 8–14.

Fuson, Karen C., Yolanda De La Cruz, Stephen T. Smith, Ana Maria Lo Cicero, Kristin Hudson, Pilar Ron, and Rebecca Steeby. "Blending the Best of the Twentieth Century to Achieve a Mathematics Equity Pedagogy in the Twenty-first Century." In *Learning Mathematics for a New Century,* 2000 Yearbook of the National Council of Teachers of Mathematics (NCTM), edited by Maurice J. Burke, pp. 197–212. Reston, Va.: NCTM, 2000.

Fuson, Karen C., and Youngshim Kwon. "Korean Children's Single-Digit Addition and Subtraction: Numbers Structured by Ten." *Journal for Research in Mathematics Education* 23 (March 1992): 148–65.

Inagaki, Kayoko, Eizo Morita, and Giyoo Hatano. "Teaching-Learning of Evaluative Criteria for Mathematical Arguments through Classroom Discourse: A Cross-National Study." *Mathematical Thinking and Learning* 1, no. 2 (1999): 93–111.

Lampert, Magdalene, and Deborah L. Ball. *Teaching, Multimedia, and Mathematics: Investigations of Real Practice.* New York: Teachers College Press, 1988.

National Council of Teachers of Mathematics (NCTM). *Principles and Standards for School Mathematics.* Reston, Va.: NCTM, 2000.

Stigler, James, and James Hiebert. *The Teaching Gap.* New York: The Free Press, 1999.

Third International Mathematics and Science Study. *Mathematics Achievement in the Primary School Years: IEA's Third International Mathematics and Science Study.* Boston: Boston College, 1997.

8

Beyond Presenting Good Problems
How a Japanese Teacher Implements a Mathematics Task

Margaret Smith

FOR several years teachers have been encouraged to use tasks that engage students in mathematical thought processes. For example, the recommendations by the National Council of Teachers of Mathematics (NCTM) in both the *Curriculum and Evaluation Standards for School Mathematics* (NCTM 1989) and the *Principles and Standards for School Mathematics* (NCTM 2000) encourage teachers and curriculum developers to use problems that go beyond practicing routine procedures to problems that help students build mathematical connections and develop and apply mathematical concepts. The expectation is that, by incorporating problems that engage students in mathematical thought processes, teachers will provide students with opportunities to conjecture, reason, and develop new mathematical ideas. Engaging students in such discussions is relatively new for both teachers and students. To help teachers learn some useful strategies, this article offers examples and discussions of ways teachers can implement problems to help students engage in mathematical thought processes.

Tasks That Help Students Make Connections

The initial release of data from the Third International Mathematics and Science Video Study (TIMSS Video Study) promoted a lot of discussion

The data used in this article were generated as part of requirements for a doctoral dissertation at the University of Delaware using video data from the TIMSS Laboratory at the University of California, Los Angeles. I wish to thank James Hiebert for his wisdom and guidance in the project, as well as James Stigler and everyone at the TIMSS Video Laboratory and LessonLab, Inc., for access to a unique and resourceful data set.

This research was supported by a grant of the American Educational Research Association, which receives funds for its AERA Grants Program from the National Center for Education Statistics and the Office of Educational Research and Improvement (U.S. Department of Education) and the National Science Foundation under NSF Grant # RED-9452861. Opinions reflect those of the author and do not necessarily reflect those of the granting agency.

about the teaching styles in Germany, Japan, and the United States (Stigler et al. 1999). After looking at the data and the sample lessons, many observed how U.S. instruction looked very procedural in nature, whereas Japanese instruction provided more opportunities for participating in the mathematical processes encouraged in the NCTM *Curriculum and Evaluation Standards* (1989) and later, the NCTM *Principles and Standards* (2000). One aspect that received a lot of attention was the types of problems students were asked to work on during the lessons. For example, the U.S. teachers appeared to be using problems that asked students to practice routine procedures, such as using formulas to calculate perimeter and area. The Japanese teachers, however, were using problems that asked students to apply and develop mathematical concepts. The tasks used in the Japanese lessons appeared to help students engage in classroom discussions that included making and testing conjectures and reasoning about important mathematical relationships.

Although using tasks designed to engage students in these mathematical thought processes is useful in developing an environment that helps students think about and discuss important mathematical ideas, research has shown that simply using these types of tasks will not promote these discussions without support (Stein, Grover, and Henningsen 1996; Stein and Lane 1996). In particular, Stein and her colleagues found that the thinking and reasoning implied by the task statement are not necessarily the thinking and reasoning students engage in while working on and talking about the task.

Reflecting on some of the differences in instruction from the TIMSS Video Study as well as the work by Stein and her colleagues, I became interested in understanding the kinds of mathematics tasks U.S. and Japanese teachers provided students as well as how these tasks were completed during the videotaped lessons. An analysis of a subset of lessons from the TIMSS Video Study, showed that although almost 40 percent of the tasks presented by U.S. teachers were designed to engage students in making connections, only about 5 percent of these tasks were implemented in a way that publicly discussed these mathematical connections. Instead, these tasks were typically completed by applying a given procedure (Smith 2000). Many of the Japanese teachers, however, were able to help students make connections by strategically encouraging students to use mathematically rich discussions: about 63 percent of the tasks Japanese teachers presented were designed to encourage students to make connections and 35 percent of these tasks were publicly completed in a way that promoted making those connections (Smith 2000). The large difference in how teachers implemented tasks in these two countries suggests that it would be helpful to look at how the Japanese teachers support tasks in a way that helps students make connections.

The next section includes two illustrations of lessons incorporating problems designed to engage students in rich mathematical discussions. One lesson describes how a U. S. teacher tried to introduce an engaging and meaningful mathematics task. The other describes how a Japanese teacher used a similar problem in her lesson. The discussion looks at some specific aspects of each teacher's implementation to help teachers think about their own practice. The differences represent some important features for implementing problems that help students make mathematical connections. The reader should note that neither the U.S. nor Japanese illustrations occurred in any one lesson: rather, I have consolidated several of the TIMSS Video findings into the two scenarios in this article. Although the stories as presented are fiction, in each respective instance they are built from the TIMSS Video data representing U.S. and Japanese teaching.

TWO TEACHERS' LESSONS

Mrs. Jones's Problem-Solving Lesson

Mrs. Jones and Mrs. Peterson attended an exciting workshop last week, where they learned about choosing good problem-solving tasks for their students. In one of the books they reviewed, they found the prom dress task (fig. 8.1) and knew it was perfect. With the "big dance" only a month away, the context was ideal. Looking more closely at the mathematics, they thought that it tied in nicely with the units on writing equations and graphing lines they had recently covered and would help students make a transition into linear systems of inequalities. They were both certain their students would generate great mathematical ideas.

> A few days ago, Veronica and Caroline were both asked to the prom. That night, they went out to shop for dresses. As they were flipping through the racks, they each found the perfect dress, which cost $80. When they showed each other their dresses, they realized they both wanted the same dress! Neither of them had enough that night, but each went home and devised a savings plan to buy the dress. Veronica put $20 aside *that night* and has been putting aside an additional $5 a day, since then. Caroline put aside $8 *the day after* they saw the dress and has put in the same amount every day since.
>
> Today, their friend Heather asks each girl how much she has saved for the dress. She says, "Wow! Caroline has more money saved." How many days has it been since Veronica and Caroline began saving?

Fig. 8.1. Prom Dress task

After assigning the problem to the students, Mrs. Jones wanted to look at how the students were tackling the problem. As she walked around the class, she noticed that most students were having trouble just getting started. To help them, she drew three columns on the chalkboard: the left column was labeled "Days Saving for Dress," the middle was labeled "Veronica," and the right was labeled "Caroline." She then asked,

Mrs. Jones: How much money has each girl saved on the first day?

Karen: Veronica has $25 because she had 20 to start with, and Caroline has $8 because she had nothing to start with.

Mrs. Jones: Okay, good. How about the second day?

Sean: 30 and 16.

Mrs. Jones: Can you explain that?

Sean: Veronica has $30, and Caroline has $16.

Mrs. Jones: Okay, now find how many days since they went shopping when Caroline has more money.

The students returned to work, and Mrs. Jones continued to walk around the room. As some students finished, she asked them to find another way to solve the problem. She later called the students together to discuss their solutions. One student filled in the table up to the seventh day and said, "On the seventh day Caroline has $56, but Veronica has only $55, so it was the seventh day." Mrs. Jones asked if there were any other solutions. No one offered one. She then asked if it had to be the seventh day or if it could be another day. Another student replied, "Well, it is the seventh day, because that is the day she has more money!" The class then began to stir, insisting it was the seventh day. Just then the bell rang.

After class, Mrs. Peterson and Mrs. Jones discussed how students worked on the problems. Mrs. Jones noted, "Class did not go very well. The only method the students used was to substitute numbers. I just couldn't get my students to think of any other methods. Also, they all think that the seventh day is the only day Caroline has more money, not just the first of many days when Caroline is ahead." Mrs. Peterson said she had an equally unsuccessful experience and commented, "We worked so hard in learning how to pick out good problems. How are we supposed to help students generate all the great mathematical ideas in the problem?"

The issue Mrs. Jones and Mrs. Peterson raise is all too familiar: There are many resources to help select problems designed to engage students in making mathematical connections — for example, Smith and Stein (1998) and Stein et al. (2000). However, the question most teachers face is how to use these mathematically rich and engaging tasks effectively. In the next section, I present an illustration of a Japanese teacher who used a similar task in a way that elicited important mathematical connections. As in the U.S. lesson, students are asked to generate mathematical ideas, building an understanding of linear systems with inequalities. In this Japanese lesson, however, the teacher tries to use some strategies to help students formalize their mathematical ideas. Comparing the example above with the way some Japanese teachers help students develop mathematical ideas through problem solving can provide teachers with some tools they can use to help promote mathematical communication.

A Japanese Story: Mrs. Hamada's Class

After greeting the class, Mrs. Hamada asked the students to settle down and get their mathematics books out while she put the gumball task (fig. 8.2) on the board. When the class was ready, she had a student read the problem aloud and then drew two rectangles on the board. One rectangle was labeled "Ken," and the other was labeled "Brother." The students then counted out 18 pink circles and then recounted the pink circles by tens to represent that Ken was starting with 180 pieces of gum. The students then counted out 24 yellow circles and then recounted the circles by fives to show that his brother was starting with 120 pieces of gum. Mrs. Hamada then asked the students to think about the problem and try to find a solution.

> Ken and his brother enjoy chewing gum. One day, the boys go to the candy store and buy several packages of gum. Ken bought 18 ten-piece packages of gum, and his brother bought 24 five-piece packages of gum.
>
> Every day, each of the boys finishes one whole pack of gum. One day, they looked at how much gum each boy had. Ken noticed that his brother had more pieces of gum than he had. How many days has it been since the boys bought the gum?

Fig. 8.2. Gumball task

The students worked on this task while Mrs. Hamada went around the room to see what the students were doing. She noticed that many students had not described their solution methods very well and commented, "Be sure to explain your answer in a way that someone else could understand what you have done." About halfway through the class, Mrs. Hamada asked her students to present their solutions to the class.

Group 1: (*The students go up to the board and draw a third rectangle and moved one pink [10-piece package] and one yellow circle [5-piece package] from the two boys' boxes into the third rectangle, as in figure 8.3 – Group 1. Continuing this action, the students explained their work.*) We drew a box here to throw away the packages of gum they chewed. On the first day, Ken puts in one 10-piece package of gum and his brother puts in one 5-piece package of gum. At the end of the first day, Ken has 170 pieces of gum and his brother has 115 pieces of gum. We kept doing this until his brother had more pieces of gum, then we counted the packages of gum (*the student points to pairings of circles that represented how many days had passed*)… 12 days.

Mrs. Hamada: And how do you know the number of days from the number of circles?

Group 1: Because one circle is thrown out each day.

The teacher summed up their work, saying, "You took one circle from each boy, counting down by tens for Ken and by fives for his brother until his brother had more gum. This is good, but it could take a long time when the numbers get bigger. Did anyone find an easier way than this?"

Group 2: (*The students put the table in figure 8.3 – Group 2 on the chalk board.*) After the first day, Ken has 170 pieces of gum and his brother has 115. Then we kept doing this until the thirteenth day when his brother had more pieces of gum.

Mrs. Hamada: Okay, so you found out how many pieces of gum each boy had on the different days and showed it in this organized way. I think this would still take a long time if the numbers were bigger.

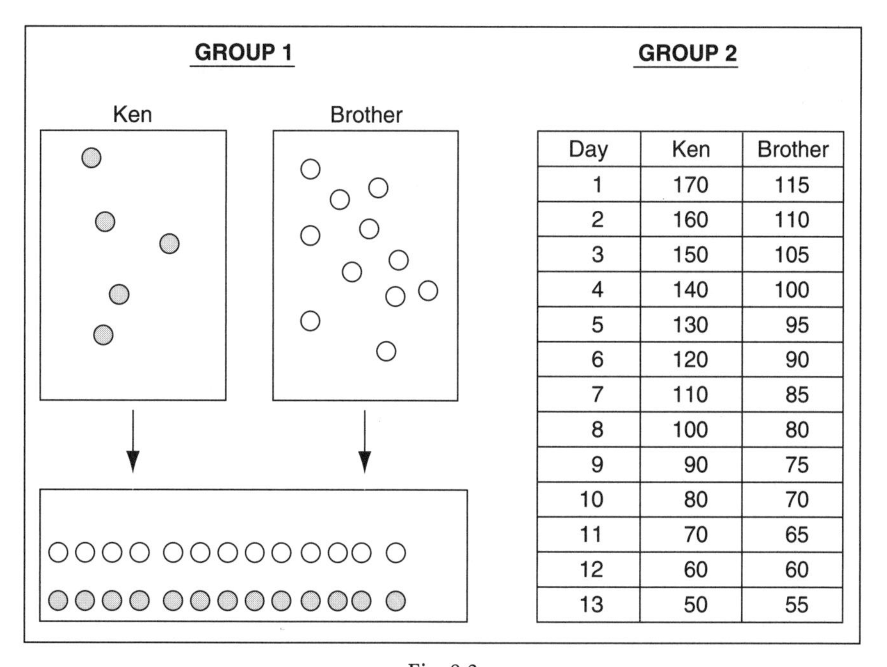

Day	Ken	Brother
1	170	115
2	160	110
3	150	105
4	140	100
5	130	95
6	120	90
7	110	85
8	100	80
9	90	75
10	80	70
11	70	65
12	60	60
13	50	55

Fig. 8.3

It appeared as if Mrs. Hamada knew of another method she wanted her students to think about; she commented, "Now I wonder if any of you thought of a way to show how many pieces of gum each boy had every day. Many of you may not have thought of this way we will do, but that is okay, we will try it anyway. I would like you to add some columns to Group 2's table like this (fig. 8.4—Before) and think of an equation Group 2 might have used to find out how many pieces of gum each boy had. What would day 1 look like?" One student from Group 2 responded, "We took 10 away from 180 for Ken and 5 away from 120 for his brother." Mrs. Hamada filled this information in on the first line and asked the students to continue.

The students worked on this task at their seats. As Mrs. Hamada walked around the room, she noticed some students were confused. She said to the class, "Okay, now stop for a minute. What I want you to look at is the two numbers: How many pieces of gum Ken has and how many pieces of gum his brother has on each day and decide the computation you used to get that number. If you find one way to show this, think about if there are any ways that may be easier." The students continued working, and the teacher had two students put their work on the board (fig. 8.4 – After). Once their work was up, Mrs. Hamada asked students to consider which equation could help them "if the number got really big." The class decided that Student 2's equation was most generalizable because "all you need to know is how many days so you can multiply it by how many pieces of gum are in each package, ten or five." There wasn't enough time to continue, so Mrs. Hamada asked the students to think of a more general way they could write how much gum each boy has on any given day.

DISCUSSION

There are many interesting ideas to talk about when comparing the Japanese teacher's use of the gumball task with the U.S. teacher's use of the prom dress task. The first is the similarity: Both teachers wanted students to use an interesting and meaningful problem to get their students to generate a mathematical idea. These teachers both provided problems in which students can generate mathematical concepts related to systems of linear inequalities. However, the teachers in these two classes differed in how they were able to help students work through the tasks. Four of the important differences follow.

- The Japanese teacher modeled the problem situation, whereas the U.S. teacher modeled a solution method.

The two teachers focused students' attentions on different aspects of the problem. Mrs. Jones tried to help students get started by offering a possible solution method, but it appears to have focused their attention on just the one particular solution method, possibly contributing to their unwillingness

BEFORE

Day	Equation	Ken	Equation	Brother
1		170		115
2		160		110
3		150		105
4		140		100
5		130		95
6		120		90
7		110		85
8		100		80
9		90		75
10		80		70
11		70		65
12		60		60
13		50		55

AFTER

Day	Student 1	Student 2	Ken	Student 1	Student 2	Brother
1	180 – 10 = 170	180 – 10 = 170	170	120 – 5 = 115	120 – 5 = 115	115
2	170 – 10 = 160	180 – 20 = 160	160	115 – 5 = 110	120 – 10 = 110	110
3	160 – 10 = 150	180 – 30 = 150	150	110 – 5 = 105	120 – 15 = 105	105
4	150 – 10 = 140	180 – 40 = 140	140	105 – 5 = 100	120 – 20 = 100	100
5	140 – 10 = 130	180 – 50 = 130	130	100 – 5 = 95	120 – 25 = 95	95
6	130 – 10 = 120	180 – 60 = 120	120	95 – 5 = 90	120 – 30 = 90	90
7	120 – 10 = 110	180 – 70 = 110	110	90 – 5 = 85	120 – 35 = 85	85
8	110 – 10 = 100	180 – 80 = 100	100	85 – 5 = 80	120 – 40 = 80	80
9	100 – 10 = 90	180 – 90 = 90	90	80 – 5 = 75	120 – 45 = 75	75
10	90 – 10 = 80	180 – 100 = 80	80	75 – 5 = 70	120 – 50 = 70	70
11	80 – 10 = 70	180 – 110 = 70	70	70 – 5 = 65	120 – 55 = 65	65
12	70 – 10 = 60	180 – 120 = 60	60	65 – 5 = 60	120 – 60 = 60	60
13	60 – 10 = 50	180 – 130 = 50	50	60 – 5 = 55	120 – 65 = 55	55

Fig. 8.4

to provide additional solutions. Mrs. Hamada also helped her students get started: In her lesson, this meant having students model the problem by showing the number of packages of gum and relating that to the number of pieces of gum each boy had at the start. This enabled her to illustrate the situation before any actions have been taken rather than setting up the information in the problem in a way that suggested a specific method. Mrs. Hamada helped students ground their understanding in the problematic situation by helping them visualize the context and then asking them to find ways to resolve it.

- The Japanese students gave more detail in their explanations. When this was not spontaneous, the Japanese teacher specifically probed students to give more detailed and connected explanations.

Both Mrs. Jones and Mrs. Hamada asked students to explain at some point, but what counted as an explanation differed in these lessons. In Mrs. Jones's class, some mathematical connections go on in the background (likely in the students' heads) rather than being made public. Simply providing an answer was not acceptable in Mrs. Hamada's class: If students did not make connections, she asked questions that linked the pieces together. For example, when asking the students to give more detailed descriptions, she used phrases such as "to explain your answer in a way that someone else could understand what you have done." This type of prompting encouraged the students to look beyond getting an answer, to describing mathematical ideas. Moreover, this level of detail in explaining one's reasoning helped to fill in the gaps for a student who may have been struggling to understand the ideas being represented.

- The Japanese teacher helped students construct another solution method.

It appeared that Mrs. Hamada had another solution method she wanted students to explore, and she wanted to help her students figure it out. From a mathematical perspective, one can assume that Mrs. Hamada was trying to get students to consider using a common variable as a first step to comparing the expressions. In an effort to help students develop this idea, Mrs. Hamada tried to get students to begin thinking of the number of pieces of gum as variable expressions. Similar to when she initially introduced the problematic situation, Mrs. Hamada offered students a way to organize their information without directing students to use one particular method. (This could be seen when the two students constructed different ways to generate equations.) By helping students to see another method, she was offering even greater opportunities for students to consider different mathematical ideas. By constructing the method after they had grappled with some of the associated counting connected to it, the students were likely to develop an understanding of how and why this solution method works.

- In the Japanese lesson, the solution methods presented were analyzed and compared.

Mrs. Hamada and Mrs. Jones valued students' solution methods. Unfortunately, Mrs. Jones's students presented only one solution method, allowing little room for developing mathematical connections across solution methods. Because Mrs. Hamada's students presented more than one solution method, she was able to have them highlight mathematical relationships. For example, the similarity of the counting method used by Group 1 and the counting-down method in the table by Group 2 were stated. As another example, Mrs. Hamada asked students to build off the ideas presented by Group 2 to construct an equation. The idea of making connections ought to be at the heart of using these types of tasks. Prompting students to give more detailed explanations and discuss their reasoning allows all students opportunities to make sense of the task, the solution methods, and the mathematical relationships involved.

FINAL REMARKS

Teaching mathematics in a way that encourages students to make connections is a challenging endeavor. Moreover, the dynamic aspects of the classroom make it difficult to provide prescriptive lists of things to do to implement mathematical tasks designed to help students make connections. The illustrations presented here are meant to help teachers recognize ways of using tasks in the classroom that help make these mathematical connections explicit. The strategies presented can help teachers orchestrate mathematics tasks in a way that promotes discussions, moving students forward in their mathematical understandings. Several features to consider have been highlighted here: modeling the problem situation, not a solution method; prompting students for more detail in their explanations; helping students construct other solution methods; and analyzing and comparing solution methods. Readers are encouraged to use these suggestions, and those identified by others (e.g., Harris et al. 2001; Henningsen and Stein 1997; Smith and Stein 1998; Stigler and Hiebert 1999), as they work to incorporate mathematical tasks designed to help students make connections.

REFERENCES

Harris, Kimberly, Robin Marcus, Karen McLaren, and James Fey. "Curriculum Materials Supporting Problem-Based Teaching." *School Science and Mathematics* 101 (October 2001): 310–18.

Henningsen, Marjorie, and Mary Kay Stein. "Mathematical Tasks and Student Cognition: Classroom-Based Factors That Support and Inhibit High-Level Mathematical Thinking and Reasoning." *Journal for Research in Mathematics Education* 28 (November 1997): 524–49.

National Council of Teachers of Mathematics (NCTM). *Curriculum and Evaluation Standards for School Mathematics.* Reston, Va.: NCTM, 1989.

———. *Principles and Standards for School Mathematics.* Reston, Va.: NCTM, 2000.

Smith, Margaret. "A Comparison of the Types of Mathematics Tasks and How They Were Completed during Eighth-Grade Mathematics Instruction in Germany, Japan, and the United States." Ph.D. diss., University of Delaware, 2000.

Smith, Margaret Schwan, and Mary Kay Stein. "Selecting and Creating Mathematical Tasks: From Research to Practice." *Mathematics Teaching in the Middle School* 3 (February 1998): 344–50.

Stein, Mary Kay, Barbara W. Grover, and Marjorie Henningsen. "Building Student Capacity for Mathematical Thinking and Reasoning: An Analysis of Mathematical Tasks Used in Reform Classrooms." *American Educational Research Journal* 33 (October 1996): 455–88.

Stein, Mary Kay, and Suzanne Lane. "Instructional Tasks and the Development of Student Capacity to Think and Reason: An Analysis of the Relationship between Teaching and Learning in a Reform Mathematics Project." *Educational Research and Evaluation* 2 (October 1996): 50–80.

Stein, Mary Kay, Margaret Schwan Smith, Marjorie Henningsen, and Edward A. Silver. *Implementing Standards-Based Mathematics Instruction: A Casebook for Professional Development.* New York: Teachers College Press, 2000.

Stigler, James W., Patrick A. Gonzales, Takako Kawanaka, Steffan Knoll, and Anna Serrano. *The TIMSS Videotape Classroom Study: Methods and Findings from an Exploratory Research Project on Eighth-Grade Mathematics Instruction in Germany, Japan, and the United States.* NCES Publication No. 1999074. Washington, D.C.: U.S. Government Printing Office, 1999.

Stigler, James W., and James Hiebert. *The Teaching Gap.* New York: Free Press, 1999.

9

Teaching to Develop Students as Learners

Debra I. Johanning

Teri Keusch

OVER the years since the publication of *Professional Standards for Teaching Mathematics* (National Council of Teachers of Mathematics [NCTM] 1991), many teachers across the United States have worked to create classrooms in which students not only learn but, more important, learn to learn. This paper explores how one sixth-grade teacher established a classroom environment that promoted the development of autonomous learners. We examine her classroom during one small window in time. In this episode, students interacted with one another to sort out why their solutions were different, and the teacher momentarily became an observer on the sidelines. It was this teacher's willingness to become an observer that allowed this moment of students' growth to occur.

This episode occurred as students were assuming greater autonomy for their own learning. But this environment for learning did not happen accidentally. It was shaped by a variety of deliberate choices and actions on the part of their teacher, Ms. Keusch. I, the first author, was an observer in Ms. Keusch's room when these episodes took place. From these observations, I begin by drawing on three specific events in her classroom, as well as on conversations Ms. Keusch and I had surrounding these events that show her deliberate efforts to cultivate her students' role as learners of mathematics. In the final episode of these events, students first showed a real interest in resolving differences in solutions and making sense of the mathematics they were studying.

This research was supported by the Connected Mathematics Project under NSF project #ESI 9986372. The views expressed here do not necessarily reflect those of NSF or the Connected Mathematics Project authors.

CREATING A LEARNING COMMUNITY: DEVELOPING THE ROLE OF THE STUDENT

Ms. Keusch teaches sixth grade in a school that took part in the initial piloting of curriculum materials developed by the Connected Mathematics Project and in the field trials of a second edition of those materials. I observed Ms. Keusch and her students over a two-month period in the fall of the school year while she piloted a revised trial version of *Bits and Pieces I* (Lappan et al. 2001–02). This unit focused on developing meanings of rational numbers and the relationships among three forms of representation—fractions, decimals, and percents.

The second problem in the unit asked students to fold fraction strips from narrow bands of paper to determine progress toward a goal at several stages in a fund-raiser (see fig. 9.1). This problem provided a context for finding different strategies for dividing up a length and then relating the parts to the whole in order to estimate fractional amounts. During the discussion of progress on Day 8, one student offered to the class that she first tried to use the thirds strip she made to measure the Day 4 thermometer. When it did not work, she tried fourths. Ms. Keusch asked how she folded the fraction strip. The student demonstrated at the overhead projector how, as opposed to folding, she used the unshaded part of the fund-raising thermometer to mark off her paper strip repeatedly into fourths to measure the goal.

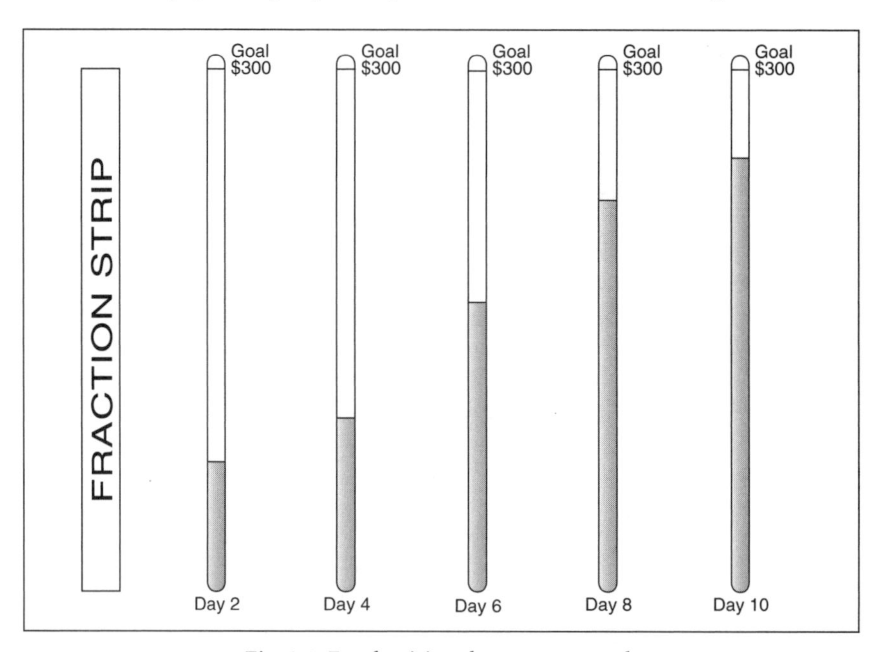

Fig. 9.1. Fund-raising thermometer task

After talking about other methods that students used to find fourths on the fraction strip, Ms. Keusch asked, "What happens when you fold a fourth in half?" Students responded, "You would get eighths." From my observations, I thought the big ideas surrounding folding strategies had come out and the students appeared ready to move on and discuss the fund-raising thermometer for Day 10. Instead, Ms. Keusch asked for other answers. "Did anyone think about the problem another way?" One student offered that he made eighths and the answer could be 6/8. Ms. Keusch drew fraction strips showing 3/4 and 6/8 on the board.

T: How are 3/4 and 6/8 different?
S1: The eighth pieces are smaller.
T: Why do we need 6 here?
S2: Since the pieces are smaller, we need more.
S3: I think 12/16 would work.
T: How could we try to get sixteenths?
S3: Cut the eighths in half. So instead of 6 eighths, we need 12 sixteenths.

Ms. Keusch's goal was for students to solve the problem posed and, from their proposed answers and strategies, to push for deeper understanding. In this example, her appeal for additional ideas prompted a student to suggest a solution that may not have been offered if the conversation had ended with folding into fourths. "I always ask, because there might be another way to think about the problem. I also knew that equivalence was the big mathematical idea in the next lesson, and I was trying to see if it would come out naturally in this problem." As a result, students were faced with resolving whether these fractions were the same. Ms. Keusch's invitation to share other ideas also sent a message to the students that they had a role in deciding when to move on in discussion and when to offer another idea to explore.

DEVELOPING NORMS FOR WHOLE-CLASS DISCUSSION

Two other explicit actions commonly taken by Ms. Keusch to develop classroom norms occur during discussions. First, when students came to the front of the class to share their reasoning, they often had difficulty expressing their thinking. In addition, there were times when they had difficulty fielding the questions raised by Ms. Keusch or the class about the presentation. When classmates posed the question, Ms. Keusch did not answer the question if the student was stuck. Rather, she reminded the student that they could ask for help from another student. "It is about them, the students, taking responsibility for their learning." Responsibility for learning was placed in the hands of the students in order to encourage them to develop behaviors that would support their ability to persevere even when they were stuck.

Second, Ms. Keusch wanted students to learn to learn from one another. She encouraged students to ask questions of one another. When students were not clear about something another student presented or said during discussion, they would ask questions about the other student's solution. But they would often direct the question to Ms. Keusch rather than to the student who was making the claim or presenting the idea. Ms. Keusch would then have them ask the student directly.

It was important that students learned to engage with one another as they tried to make sense of the mathematics. "We are in a community, and if I always restate the question or answer questions for them, the conversation is between that one student and me. If I take over, then they depend on me and do not listen and learn from each other. The goal is to have a classroom conversation, not a conversation between each student and me." Although there were times when it made sense for Ms. Keusch to support discussion by asking questions, clarifying ideas, and providing information, there were also times when the role of the student in discourse had to be made explicit to the students so they would learn to see one another as resources for learning mathematics.

Progress toward Learning How to Learn

Toward the end of the *Bits and Pieces* I unit, students were investigating percents. In a previous problem, students explored the idea that percents are useful when one wants to make meaningful comparisons. In this problem (see fig. 9.2), students worked with information provided in tables from a database about cats. Students were asked a series of questions that could be answered by manipulating data from the tables.

The questions in this problem were designed to promote moving among fractions, percents, and decimals. The number of cats in the data base totals 100, which enforces the notion that it is simple to move among different forms when the total is 100. One main goal of this problem is for students to confront a situation that is not "out of 100." Question C asks about the fraction of kittens, a group that does not have a total of 100. In addition, the discussion about changing forms allows students to consider how moving from fractions to decimals to percents where the total is out of 17 may require a different approach from what was used in question A or B, when the total was out of 100.

As students worked on the problem in small groups, it became obvious to Ms. Keusch and me that most of the students were struggling with question C. They did not realize their answers and strategies were inaccurate. The stu-

A middle school class had assembled a database on 100 cats owned by students in the class and other students in the school. In order to make sense of their data, the students carefully sorted the database and made the following tables to answer some questions they found interesting.

Jane wondered what percent of the cats were female. Tang is interested in kittens. He wants to know what percent of the cats in the database are kittens (eight months old or younger) and what percent are adults (more than eight months old). To answer their questions, they made table 1.

Table 1

	Male	Female
Kitten	10	7
Adult	36	47

Jamie's adult cat, Melissa, weighs 11 pounds, and Barb's kitten, Seymour, weighs only 1.5 pounds. They made the following table to look at male and female adult cats and kittens and their weights.

Table 2

Weight in lbs.	Male Kitten	Male Adult	Female Kitten	Female Adult
0 – 5.9	8	1	7	4
6 – 10.9		16		31
11 – 15.9	2	15		10
16 – 20		4		2

Answer the questions using tables 1 and 2 that the students made.

A. What fraction of the cats are female? What fraction of the cats are male? Write the fractions as a decimal and a percent.

B. What fraction of the cats are kittens? What fraction of the cats are adult? Write the fractions as a decimal and a percent.

C. What fraction of the kittens are male? Write the fraction as a decimal and a percent.

D. Write and solve four questions that can be answered from the data in table 2. Be sure to show each of your answers as a fraction, a decimal, and a percent.

Fig. 9.2. Cat data tables and questions

dents just assumed that the total was out of 100, as had been true in previous problems. However, one group of two boys realized that finding the total was important and this category did not add to 100. Ms. Keusch asked some students to put their solutions to the first three problems on the board. She then said, "After Lisa, Amy, and Mark put their solutions to A, B, and C on the board, if anyone has a different approach or answer, put yours on the board under theirs." While the students were putting solutions on the board, other students were finishing question D.

First, Mark put a solution to question C on the board. Next, T.J. put a different solution below Mark's (see fig. 9.3). When Mark realized that T.J.'s solution was different, he returned to his seat, reread the problem and then went to the board to ask T.J. how he got his answer. While T.J. explained his solution to Mark, a pair of girls who had been working together, Heather and Mari, noticed that their solution matched Mark's and not T.J.'s. They, too, went to the board to ask T.J. how he got his answer. As the girls finished talking with Mark and T.J., they returned to their seats, saying, "I am going to revise my answer to show the number of kittens is out of 17."

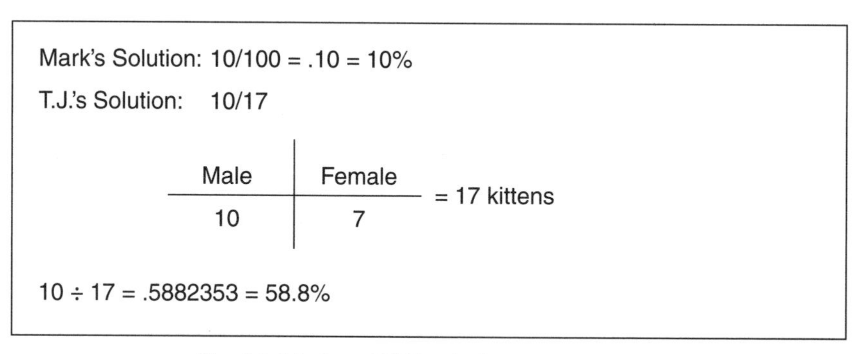

Fig. 9.3. Mark and T.J.'s solutions to question C

As this interaction ended, Ms. Keusch brought the class back together to discuss the problems on the board. For problem C, Mark and T.J. both were at the board together.

Mark: I have decided to revise mine to be like T.J.'s. When I did this [*pointing to his solution*] I did not think about how many kittens there were altogether. My answer says there were 100 kittens and 10 were male. But there were only 17 kittens.

T.J.: I used table 1 [*in the problem*] and saw that there were 10 male kittens and 7 female kittens so I knew there were 17 kittens altogether. So to find the fraction of kittens that were male I wrote 10 out of 17 and divided to find the percent.

REFLECTING ON WHAT HAPPENED

The goal of teaching is to promote students' learning. The classroom environment can promote or ignore students' making sense of the mathematics. When sense making and independent learning are the goal, teachers must expect students to explain their ideas, to justify their solutions, and to persevere when they are stuck (NCTM 1991). Mathematical tasks alone are not sufficient for effective teaching. Teachers need to decide how to organize the environment for learning by considering how students will present their work, what questions to ask, and how to support students' learning without taking over the students' process of thinking (NCTM 2000).

As I watched this classroom episode unfold, three ideas stood out. First, Mark was not flustered about being wrong and saw that T.J.'s reasoning was justified. He did not remove his answer from the board when he revised his position. Second, for students who had Mark's solution, Mark's explanation of his inaccurate reasoning was as important as T.J.'s explanation of his correct reasoning. The episode was strengthened by the collaboration between the two students as they explained how they made sense of the problem. And third, Ms. Keusch never had to elicit students' comments. Her strategy of having students put multiple solutions on the board provided an opportunity for her to facilitate a discussion, but she never had to start the conversation. These students—Mark, T.J., and the girls, Heather and Mari, who asked about Mark and T.J.'s solution—had developed the disposition to question one another and engage in a genuine conversation about the mathematics in the problem. They became the authorities. The choice that Ms. Keusch made to have the students put multiple solutions on the board was part of what allowed her to stand aside and observe as her students directed their own learning.

There were many occasions prior to this event where Ms. Keusch was both implicit and explicit in her facilitation of the learning environment. Across the two months I observed in Ms. Keusch's room, I saw her continually seek out multiple solutions during whole-class discussions and use the format of having students display multiple solutions prior to discussion. Yet, this was the first time I saw students interact independently outside their immediate, small working groups to make sense of the mathematics they were exploring.

ESTABLISHING AND MAINTAINING THE LEARNING ENVIRONMENT

The *Professional Standards for Teaching Mathematics* (NCTM 1991) proposes that teachers need to do more listening and students need to do more reasoning. Students need time to be stuck, to puzzle, to try alternative approaches, and to confer with others. Ms. Keusch explains, "When I first started teaching, I used to go around and try to resolve all the problems that came up in small-group discussion. But there was nothing left to talk about in the whole-class discussion." As we saw in the example with Mark and T.J., Ms. Keusch had developed strategies to monitor and organize participation actively so students could develop the disposition and ability to engage in mathematical activity in both large- and small-group discussion.

In addition to thinking about *her* role in discourse, Ms. Keusch actively worked to help students understand *their* role in classroom discourse.

> Students should engage in making conjectures, proposing approaches and solutions to problems, and arguing about the validity of particular claims. They should learn to verify, revise, and discard claims on the basis of mathematical evidence and use a variety of tools. Whether working in small or large groups, they should be the audience for one another's comments—that is, they should speak to one another, aiming to convince or question their peers. Above all, the discourse should be focused on making sense of mathematical ideas, on using mathematical ideas sensibly in setting up and solving problems. (NCTM 1991, p. 45)

Ms. Keusch described the development of this process as "a collection of subtle things starting when they walk in the door." Behaviors such as looking at the speaker, asking questions appropriately, talking about what you think, and listening are important to reinforce. "When I see someone make a mistake that forces the rest of us to learn more and question more, I make a big issue out of it." In addition to public acknowledgment, Ms. Keusch pulled aside students to talk with them, reinforced their asking good questions, or pushing the class's thinking. She also took students aside to talk with them about behaviors that were not helpful. "I do a lot of that early on so the tone is set, and then I do a lot of modeling all year." For Ms. Keusch, creating a classroom that supports discourse and facilitates learning is a continuous and ongoing process.

SUMMARY

Effective mathematics teaching involves deliberate and ongoing efforts by the teacher to help students understand their role as learners and doers of mathematics. "Ultimately, students must assume responsibility for their own learning. However, the teacher has the responsibility to create an environment in which students are encouraged to accept that responsibility" (NCTM 1991, p. 116). These efforts appear in many forms, ranging from explicit to implicit. At times, they even require restraint, patience, and silence on the part of the teacher to allow the moment to unfold.

As teachers work to facilitate students' learning, Ms. Keusch's teaching reminds us that classroom discourse has to be developed and maintained throughout the year. There were several things that Ms. Keusch did to help students understand their role as learners, especially when they participated in whole-class discussions. Her ideas can help all of us strengthen our classroom environment.

- Explain your expectations to students.
- Bring students into the conversation. Prompt students to offer their ideas and their reactions to other students' ideas.
- Actively look for moments when students exhibit helpful behavior. Reinforce helpful behaviors both publicly and individually.
- Address nonhelpful behavior early on.

Addressing such issues is important at the beginning of the year. In addition, a teacher must monitor and maintain classroom discourse and students' learning throughout the year.

- Prompt students to offer their ideas. "Did anyone think about the problem in a different way?"
- When students are explaining or offering a strategy, prompt them to come to the front of the class to present and defend their ideas. Reinforce those students who take the initiative to come up to speak or ask if they can.
- Prompt students to react to other students' ideas. "Does anyone have any questions they want to ask Joe about his solution?"
- When students seem unsure, subtly prompt them to address their confusion. Ask a student if they heard what Joe said and to repeat it. If they can't, direct them to ask Joe to repeat what he said. Ask students to explain another person's idea. If they struggle, suggest they call on another student for help.
- Promote sense making and independent learning by allowing students to find their own errors. When misconceptions exist, find ways to direct students to examine those misconceptions without being critical and without pointing them out explicitly.

Teachers often expect students to work together in small groups. When students are working together in small groups, they are more likely to engage in interactive discourse. The structure of small-group interaction lends itself naturally to investigation, conjecturing, and arguing about the validity of an idea. Students need help learning to take these behaviors learned in small groups into the whole-class discussion. In the example with Mark, T.J., Heather, and Mari, their actions are reflective of the norms that Ms. Keusch worked to establish during whole-group discussion. In a large-group setting, students are often used to the teacher taking the lead, and they may be less comfortable with their role in large-group discussion. Yet, it is in the large group where the important mathematical ideas are put on the table for all students to examine and make sense of. As the teacher, you want to consider your role in this environment. When should I lead? When should I redirect? When should I step back? It is important to reflect on the message that you as the teacher are sending to students about their role as learners and participants in a mathematical community. In addition, by helping students develop strategies for interacting in class discussions, we can help students develop autonomy and greater responsibility for their own learning.

REFERENCES

Lappan, Glenda, James Fey, William Fitzgerald, Susan Friel, and Elizabeth Phillips. *Bits and Pieces I: Understanding Rational Numbers, Student Edition.* Draft version. Glenview, Ill.: Prentice Hall, 2001, 2002.

National Council of Teachers of Mathematics (NCTM). *Professional Standards for Teaching Mathematics.* Reston, Va.: NCTM, 1991.

———. *Principles and Standards for School Mathematics.* Reston, Va.: NCTM, 2000.

10

Enlarging the Learning Community through Family Mathematics

Phyllis Whitin

THE teacher is responsible for creating an intellectual environment where serious mathematical thinking is the norm" (National Council of Teachers of Mathematics [NCTM] 2000, p. 18). But does the environment stop at the classroom door? How might the community of inquirers be extended to include families? These are questions that I pondered one summer, and they led me to design a family mathematics program for my fourth-grade class.

Rationale

Identifying goals was the first step in creating the program. I knew that I wanted the children to be engaged in meaningful homework that would enrich life in the classroom. I also wanted parents and guardians to better understand the important role that reasoning plays in their children's school mathematics. My goals therefore included the following:

- Demonstrate to parents and guardians the content and process of classroom mathematics through active involvement
- Build parents' and guardians' appreciation of their children's thinking
- Provide an additional audience for children to communicate mathematically
- Create opportunities for adults and children to see one another as fellow thinkers
- Build students' confidence and positive dispositions toward mathematics
- Increase time spent with worthwhile mathematical activities
- Enrich the class's collective pool of mathematical language and strategies

DEVELOPING THE CONTENT AND PROCEDURE FOR "FAMILY MATH"

In designing experiences for children, I wanted to "pique students' curiosity and draw them into mathematics" (NCTM 2000, p. 18) at home as well as at school. At the same time, I wanted to provide experiences that connected important mathematical ideas over time. I developed the following criteria for these tasks:

- Incorporates reasoning, such as forming conjectures and testing them, or identifying and describing patterns
- Is open-ended so that all learners can be challenged
- Involves manipulatives or visual forms of representation
- Provides experience with important mathematical content

I planned for a home-school mathematics experience about once a month throughout the school year. Here is the general structure I followed:

- Inform parents and guardians of the goals of family math
- Involve students in a mathematical investigation, including time for collaboration and communication of findings
- Develop procedures for children to conduct similar investigations at home, having the children specifically document their families' reasoning and strategies
- Allow several days for the children to work at home so that each family's schedule could be accommodated
- Provide time for children to share their findings with the class
- Have children reflect on the similarities and differences among their strategies and reasoning, and those of their families

I found another aspect of this process shortly after the year began. As I will describe in the stories that follow, the children suggested extensions of the classroom investigations I had planned. Once I saw their extension, I wanted to incorporate their personal interests into home-school experiences. I would now modify the list above to reflect the extension of planned tasks.

LAYING THE FOUNDATION: IN-CLASS EXPERIENCES

The first mathematical topic of the year involved classification activities with attribute blocks (fig. 10.1). Past experience had shown me that many students came to fourth grade with a narrow perception of mathematics as being only computation. Work with attribute blocks could widen this view by emphasizing mathematical thinking and reasoning. I also could present activities through nonthreatening games. (The attribute blocks activities described in this article are based on *Attribute Games and Problems* [Elementary Science Study 1968].) The experiences would also introduce Venn diagrams as a form of representation that could be used in different ways throughout the year.

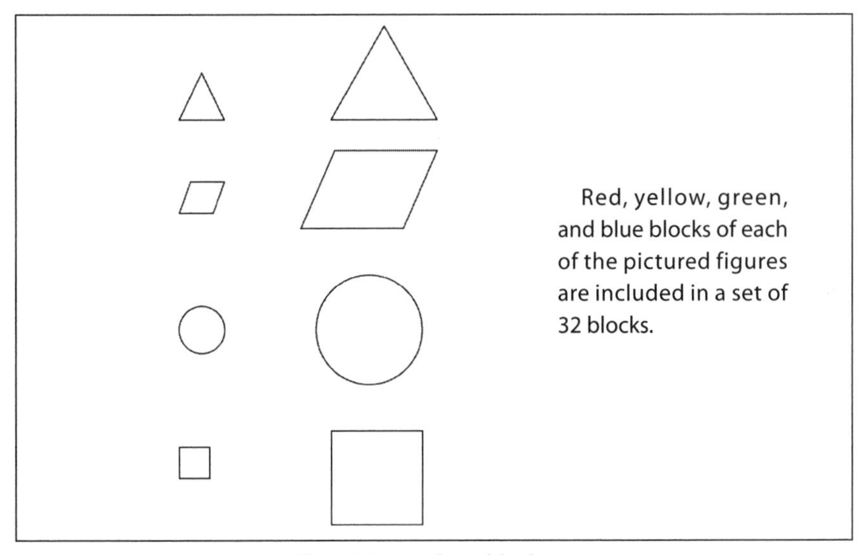

Red, yellow, green, and blue blocks of each of the pictured figures are included in a set of 32 blocks.

Fig. 10.1. Attribute block set

Since the students were unfamiliar with attribute blocks, I began with a deductive reasoning game called "the mystery bag." In this game, the teacher places the entire set of blocks in a bag. She informs the children that the bag contains some objects, and their task is to use the information from the first object to make hypotheses about the remaining contents. The children ask yes-no questions to test their hypotheses. The teacher removes a piece for each of the "yes" answers. For example, the teacher might begin by taking a large, yellow square from the bag. A child might ask, "Is there another square in the bag?" in which instance the teacher could reveal another large square (but a second color) or a small square (yellow or another color). If the child asks if there is another yellow shape in the bag, the teacher might remove the small yellow square, or a yellow circle, triangle, or parallelogram (either size). Throughout the process of asking questions and revealing pieces, the children use deductive reasoning to find the attributes that apply to this set of pieces. A value of this activity is that a negative response ("No, there is not a ____ in the bag") contributes valuable information.

After the class played "the mystery bag" and all the pieces were in view, I asked the children to describe in their journals what was going through their minds during the game. Justin, for example, related that at first he had thought that the bag contained only a variety of large shapes. With this hypothesis in mind, he asked if there was a hexagon in the bag, but the answer was negative. He described his reasoning at this point in the game: "When 'hexagon' didn't work, I knew I should try something else like 'little'" (another attribute). Such reflective talk and writing was an important step in developing a community that valued reasoning, and it served as a rehearsal for later work with families.

I also taught the children the game "guess my piece," in which the entire set of blocks is placed in view of two partners. One player selects a piece mentally, and the partner asks a series of questions about the piece's attributes (size, shape, color) in order to identify it (e.g., "Is your piece large?" "Does it have four sides?" and so on). Finally, I introduced Venn diagrams through a game called "guess my rule." Here the teacher sorts the blocks in two intersecting yarn loops (a Venn diagram) according to certain attributes. An example is shown in figure 10.2. The students' task is to study the members of each set and identify the attributes that pertain to each set. In this example, the children would label one loop "blue"; the second, "triangles"; and the intersection, "blue triangles." Once familiar with this game, the children created similar puzzles for one another and solved increasingly complex puzzles that I drew on the board to begin each day. Children described their reasoning orally as well as by writing and drawing in their journals.

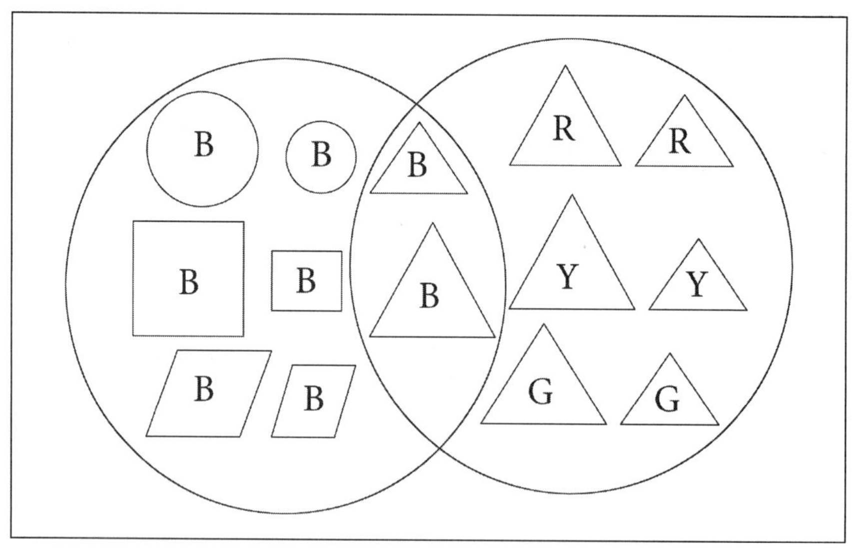

Fig. 10.2. An example of a "guess my rule" puzzle. (Letters are the first initials of colors: blue [B], red [R], yellow [Y], or green [G].)

I was pleased with the students' work with attribute blocks as a first experience for the year. With varying degrees of complexity, each child successfully constructed "loop puzzles" (Venn diagrams), which met my criteria for challenging all students. I thought that similar experiences would work well to begin my program of family mathematics. The children would lead the games at home, which could serve to increase their confidence. Having the children observe and report on their families' strategic thinking would give them opportunities to reflect further on their own reasoning. Finally, the games could open doors for additional talk about mathematics at home.

The First Home-School Investigation

The children role-played how they could teach these two games to a family member. I used a die-cutting machine to create a set of paper attribute blocks for each child, made loops of two different colors of yarn, and enclosed all materials in individual zip-lock bags. I included instructions for the student, along with a letter to the families (see figs. 10.3 and 10.4).

1. Take your shape pieces home, and play the "guess my piece" game with someone in your family. Play this game several times with this person. Ask how the person figured out which piece it was, and write down what was said. You may want to draw a picture to help you explain that person's thinking.
2. Now make up two loop puzzles for your partner. Ask your partner how he or she figured out these puzzles ("guess my rule"). Write down what was said. Make a picture of your puzzle, too.
3. You may play other games such as "missing piece" or "mystery bag" if you wish.

Fig. 10.3. Instructions for students

After everyone had completed the activities at home, the children discussed their experiences. Their comments dealt with mathematical dispositions, such as persistence, as well as with strategic thinking. Molly, like other children, reported about her mother's persistence (in this instance, for "mystery bag:): "[My mom] said that it was hard at first, but then, when she started to get the hang of it, she got how to do it." Jeff related his observations of his older brother when he told him to put all squares in one loop and to put all the small ones in the other. Jeff's brother studied the pieces silently for a few moments. Suddenly he said, "Ahhhh," because he "figured that there was a patch (intersection) in the middle." Both these students expressed interest

Dear Parents/Guardians:

On several occasions during the year, your child will bring home an activity to share with a family member. Today your child has brought home the first one. Each activity will be one that we have completed at school. The purpose of Family Math is to give children an opportunity to share a school experience with a family member (including older siblings), adult friend, or caregiver. By talking about the activities together, you will learn more about your child's thinking strategies, and your child will learn from you. Most of the activities are puzzles and can be solved or explored in many different ways. The requirements for each assignment should not be overly time-consuming, but the activities will be open-ended so that you can spend more time if you wish.

I will always allow at least two days for a Family Math assignment to accommodate busy family schedules. I want these experiences to be enjoyable for everyone!

Fig. 10.4. Letter to families

in the fact that their family members faced initial puzzlement in similar ways that they had when I introduced the same activities in class. Other children described their partner's logic. Some children were interested that their families shared strategic thinking similar to their own. For example, Courtney wrote that in "guess my piece," her mother asked if the pieces were "big or small, does it have four sides, and then my mom asked about the shape, then the colors. So my mom sort of used the same strategy as I did when I played at school." The activity thus addressed the goal of highlighting both attitudinal and strategic similarities across ages.

The Children Inspire the Second Family Experience

It is important to note that the children used Venn diagrams in a variety of contexts after I introduced the attribute block games. For example, in language arts the children interviewed one another and created diagrams to record their individual and shared interests. They used their diagrams as an organizational tool to write descriptions of their partners. Soon the children were experimenting with using Venn diagrams in other ways. One of their inventions grew out of an experience with plotting multiples of different numbers on hundred charts, and it inspired the next family experience.

While studying the pattern of shared multiples on one chart, Madison compared the intersections with those on Venn diagrams (fig. 10.5).

	2	3	**4**	5	**6**	7	**8**	9	**10**
11	**12**	13	**14**	15	**16**	17	**18**	19	**20**
21	**22**	23	**24**	25	**26**	27	**28**	29	**30**

Fig 10.5. A portion of a hundred-chart identifying multiples of 2 (bold) and 3 (shaded)

Intrigued with her idea, I adapted the "guess my rule" game to this numerical context. For example, I gave the children a diagram of numbers in two intersecting loops and asked them to figure out the rules that governed it (fig. 10.6).

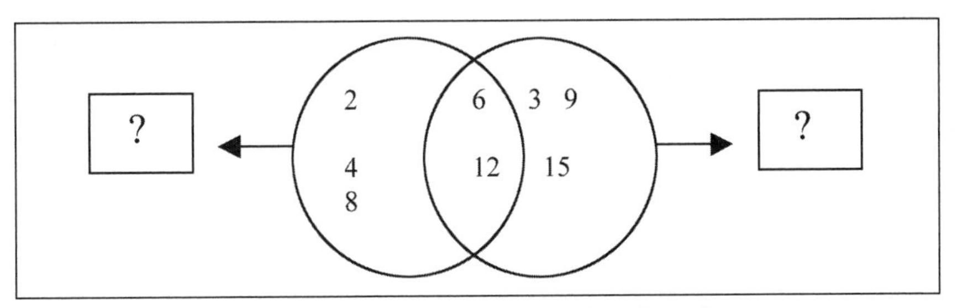

Fig 10.6. Venn diagram of multiples of 2 and 3

The children discussed their processes of reasoning and recorded their ideas in their journals, just as they had done for the geometric shape puzzles (e.g., trial and error of different times tables). Soon children were creating similar puzzles for their classmates to solve. As the children's interest in these representations increased, I decided to use this idea for the next Family Math experience. By using puzzles with which they were very familiar, the children could once more focus their attention on their families' thinking. I designed a sheet that showed three Venn diagrams. Together the children filled in numbers according to the following specifications:

- Multiples of 2 and multiples of 3 (intersecting sets)
- Multiples of 2 and multiples of 5 (intersecting sets)
- Multiples of 4 and multiples of 8 (set within a set)

The children left the labels for the puzzles blank as they had done with the class examples. The instructions for homework included "Describe how your family partner figured out the rules for the numbers in the diagrams. What did your family partner find interesting about the sets of numbers?"

In the class sharing session, the children described the family members' processes of thinking. Once again, some children commented on their families' perseverance (e.g., "She took awhile but she got it"). Others described strategies, many of which were quite similar to those that the students had described in class. One strategy was to look at the intersection first, as Molly reported, "My mom just looked at the middle section to see what numbers there were in common." In other words, 6 is a factor of 12 in figure 10.6. This relationship gave clues for solving the puzzle. Will's mother said she used "[T]rial and error until the pattern fit the scheme." She hypothesized a rule for a pattern and then tested that pattern against the members of the set. Although the children were already aware that all members of a given set have to follow the same pattern, it was beneficial for them to hear that adults checked their ideas in similar ways.

Other family members expressed their strategies in new ways for the children to consider. For example, Jessie's father expressed his strategy very systematically and succinctly. His brevity and efficiency impressed his daughter. She said, "My dad is smart! He said that whatever the lowest number is, that is what you are counting by!" In other words, he found the lowest number in each of the large circles (2 and 3 in fig. 10.6), reasoned that this number began a pattern, and tested multiples to confirm his hypothesis. This example and others demonstrated alternative ways for the children to express rules for a pattern. In addition, the children's reports included a rich range of ways to describe the set of multiples of 8 within the set of multiples of 4 (fig. 10.7).

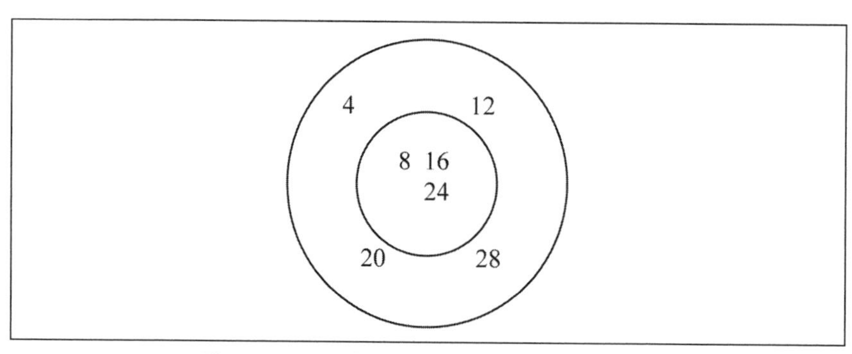

Fig 10.7. Venn diagram of multiples of 4 and 8

Below are some examples of the students' written statements, with my comments in italics.

- Count by 4s in the outer circle and count by 4s and 8s in the inner circle. All numbers are even. *(Describes multiplication as repeated addition; highlights common multiples, even numbers)*

- My mom found the middle numbers [the intersection] interesting because [those] numbers belonged with each set of numbers. *(Describes the representation)*
- 4 is doubled to 8. 8 is doubled to 16. *(Shows the doubling relationship between each member of the outer ring with its corresponding member in the inner circle)*

The last child and his mother became intrigued with the doubling idea and extended the representation to include four concentric circles with one member each: 4, 8, 16, 32, which he was able to share with the class (fig. 10.8).

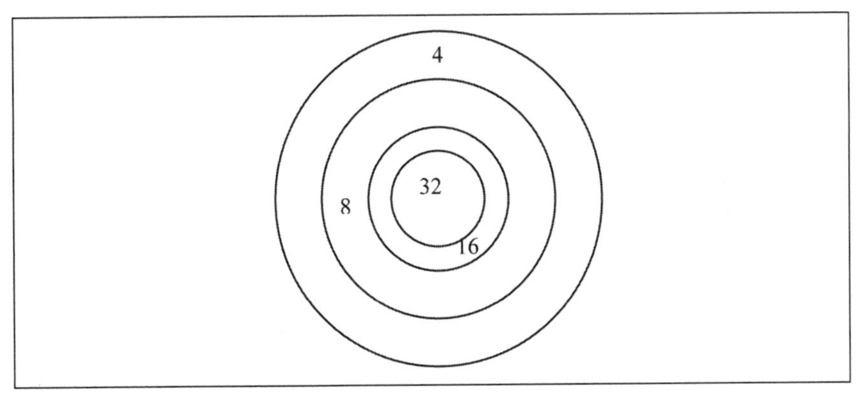

Fig. 10.8. A family-generated extension of figure 10.7

His example highlighted for me as a teacher the importance of designing tasks that were open-ended. The prompt "What did your family partner find interesting?" allowed all children (and their families) to be successful yet challenged.

All the descriptions from home definitely enlarged the children's familiarity with factors and multiples. Each child had the opportunity to discuss the diagrams one-on-one with an adult. I also suspected that reporting on their parents' insights built the confidence of the more shy students in the class, particularly for a child who received special education services. In addition, providing time for the children to share their experiences exposed them all to a variety of ways to express mathematical relationships, such as common multiples and division as the inverse of multiplication.

Reflections on the Program

These two experiences laid the foundation for continued family investigations throughout the year. In some assignments, the children continued to report on their partner's thinking; in others, the two worked together to investigate patterns (Whitin and Whitin 2002). Table 10.1 summarizes the framework that guided the development of these family mathematics experiences. At the conclusion of the year, I asked both the students and parents to evaluate the program by commenting on what they enjoyed, what they discovered or were surprised by, and how they would change the program.

Comments from the first two prompts highlighted benefits of the project.

Students:

- What I liked about Family Math is that you get to really see how smart your parents are.
- I enjoyed seeing my family make predictions.
- I found a new way to do the problem.
- [I liked that] you spend time with your parents.

Parents and Guardians:

- [I found] that she was able to solve problems through her own ingenuity.
- I learned a lot about Jenna's ability to reason. I enjoyed spending time with her learning something new.
- I enjoyed how persistent Jeff was until he completed all the problems.
- I enjoyed having to sit and think things through with Samantha.
- Cody thinks through situations, as to what the situation will be, not just the moment.
- [I was surprised] seeing my son might solve a problem differently from me. He and I both think our way is the only way to solve a problem.

These comments indicated to me that I had met some of my goals. The children were comfortable talking about mathematical ideas with their parents, and they increased their confidence by casting themselves in the role of teacher and observer. I also witnessed that shy students contributed more readily in class as a result of Family Math experiences. The parents' comments overwhelmingly showed an appreciation for their children's mathematical thinking. I was particularly surprised by the final quote in the list above. Both father and son learned to "listen to and understand conjectures and explanations offered by [others]" (NCTM 2000, p. 57). Finally, observing the children in class during sharing sessions helped me assess their confidence as well as their ease in explaining mathematical ideas. As a teacher, I could often capitalize on a family member's idea to extend all the children's mathematical learning.

LOOKING AHEAD: STRENGTHENING THE PROGRAM

Comments from students and parents helped me make plans for strengthening the program. One student, for example, suggested giving

Table 10.1

Planning for Family Mathematics Activities

Criteria for Family Mathematics Activities

- Is open-ended; extensions can challenge all learners
- Involves reasoning
- Can be solved through multiple strategies
- Focuses on reasoning and communication
- Involves manipulatives or visual forms of representation
- Provides experience with important mathematical content

(Note: Not all criteria will apply to every activity.)

Before a Task Goes Home

- Students are familiar, competent, and confident with the task, so that they can focus their attention on their family members' thinking.
- Students are experienced in describing and justifying their reasoning, in oral, written, and graphic forms (mathematics journals, seminars, forums).
- Parents are informed about the goals of the family mathematics program.

Activity Topics to Consider

- Data and data analysis
- Probability
- Geometry and geometric design
- Patterns (both numerical and geometric), functions
- Logic
- Appreciation of math as a tool (e.g., history of measurement, numeration systems)

Sample Questions to Include in Family Mathematics Homework Guidelines

- What did your family member notice about … (e.g., pattern, geometric shapes)?
- What did your family member find interesting about …?
- Describe something that surprised your family member about….
- Describe how your family member figured out … (e.g., the rules for this pattern, the numeration system).
- How was your family member's strategy like yours? Different from yours?

After the Family Mathematics Experience

- Provide a time for class debriefing.
- Communicate to parents the impact of the activity on the mathematical thinking of the students (newsletter, parent nights, notes with report cards, or other means).

"more pictures to help the family member." Some parents' comments also showed me that they were unaware of the impact that the home activities had on the class's developing knowledge and attitudes throughout the year. I therefore planned to make these benefits more explicit through comments in children's report card folders and in the class newsletter. For example, I could write a short report on the sharing sessions that followed each of the home activities, citing some of the parents' comments with their permission (Mills, O'Keefe, and Whitin 1996). Giving a workshop about the *Principles and Standards* (NCTM 2000) or creating a video or Web-based slide show might better communicate to parents my goals of developing reasoning and communication.

A family mathematics program has a rich potential for providing an intellectual environment that values reasoning and promotes understanding. Students develop confidence and positive dispositions toward mathematics when they play the role of leader and investigator. Students gain additional experience engaging in mathematical discourse, and families learn ways to encourage the children's reasoning. When students report their findings to their peers, all class members learn new strategies and forms of expression. In these ways teachers can work to enrich the mathematical experience for their students and to extend mathematical discourse beyond the classroom door.

REFERENCES

Elementary Science Study. *Attribute Games and Problems.* New York: McGraw-Hill, 1968.

Mills, Heidi, Timothy O'Keefe, and David J. Whitin. *Mathematics in the Making: Authoring Ideas in the Primary Classroom.* Portsmouth, N.H.: Heinemann, 1996.

National Council of Teachers of Mathematics (NCTM). *Principles and Standards for School Mathematics.* Reston, Va.: NCTM, 2000.

Whitin, Phyllis, and David J. Whitin. "Promoting Communication in the Mathematics Classroom." *Teaching Children Mathematics* 9 (December 2002): 205–11.

11

Building a Mathematical Community through Problem Posing

David J. Whitin

An important aspect of effective teaching is the strategy of problem posing (Brown and Walter 1983 [1990]). It is a strategy that invites learners to look closely at a mathematical problem or story, and then modify that situation in ways that interest them. Problem posing can have a profoundly positive effect on the norms of a classroom environment. It is a strategy that builds a spirit of intellectual excitement and adventure by legitimizing asking questions and freeing learners from the one-answer syndrome (Brown and Walter 1990, p. 5). In this environment, students are expected to question assumptions, challenge ideas, and extend mathematical investigations. Problem posing can encourage this kind of investigative climate.

Three Problem-Posing Stories

Let's start by examining three stories of learners engaged in posing their own questions. Then we'll step back and discuss the teacher's role in employing this strategy to stretch their students' mathematical thinking.

Story #1: Observing Garages

Susan Kelly was driving in the car one day, accompanied by her five-year-old son Jacob. As they passed a large house, Susan remarked to her son, "Look how big that house is. It has a three-car garage!" Jacob thought about his mother's comment for a moment and then replied, "They must have twelve people in that house." Susan asked him how he knew that, and he reasoned, "Well, cars have four seats, so I added up the three fours, and that is twelve." However, Jacob was not finished with this problem. A few minutes later he suggested another alternative: "You know, there could be fifteen people in that family if they squished one more person in the cars." Susan thought that since their own car had four seats, Jacob's first calculation was probably based on his personal experience with cars. She was amazed that he was able to calculate the total and even more surprised that he could pose an additional problem for himself to solve.

Story #2: Dividing Erasers

Second-grade teacher Paul Spinella created word problems for his students that were related to the school store. His students often purchased paper, pencils, and erasers at the store, and he thought that was a meaningful context for solving a variety of problems. On this particular occasion, he challenged the children to figure how three children could share 20 erasers. One child suggested that two children get 7 erasers and the third child get 6. Others disagreed and said that solution would not be fair. They finally agreed that each child ought to receive 6 erasers, with the remaining 2 erasers given to the teacher. However, the problem did not end here. The children had become intrigued with trying to divide that set of erasers evenly and wondered what numbers of erasers could be divided evenly, with no remainders. Paul described the problems they posed for themselves: "At first they thought only even numbers worked, but Lucas pointed out that 9 worked ($3 + 3 + 3 = 9$). Then they thought only numbers with 9 in them would work, but then they found that 19 did not work, ($6 + 6 + 7 = 19$). Alicia did not think 90 would work, but Lucas showed her that there were three 30s in 90. They seemed to enjoy figuring all this out."

Story #3: Finding Squares in Rectangles

During an undergraduate mathematics methods class, students were given square tiles and asked to build all possible rectangles for each of the numbers from 1 to 24. One student, Sharon, discovered that one solution for 8 tiles was a 2×4 rectangle. She then remarked, "Look, I see two squares inside this rectangle (pointing to two squares that were 2×2)." The teacher, David, responded, "That's very interesting. I wonder what other rectangles could be broken into squares." Sharon and her partner were fascinated with the idea and began working. After some investigating, they found the following pattern:

2×4 makes two 2×2 squares

3×6 makes two 3×3 squares

4×8 makes two 4×4 squares

Sharon and her partner even wondered if they could find different sized squares inside a given rectangle.

REFLECTING ON THE STORIES

Despite the differences in ages, all these learners were demonstrating important mathematical thinking. In each situation the teacher played an important role in fostering these explorations. Susan offered a mathematical observation about garages, and then later asked Jacob to explain his thinking about twelve people in the family. Having him reflect on the process may have prompted him to suggest a further extension later. Paul capitalized on his student's inquisitiveness about numbers divisible by 3 by challenging them to investigate other numbers. David encouraged Sharon to pursue her initial observation further by trying to find additional examples. Thus, teachers play an important role in supporting a problem-posing stance by sharing their own mathematical observations, asking children to explain their thinking and challenging them to find other examples to extend their initial discovery.

The strategy of problem posing also promotes a classroom environment that builds essential mathematical attitudes. For instance, in all three stories the learners demonstrated the important disposition of risk taking by posing problems for which they did not have ready answers. Teachers can promote this important attitude by encouraging learners to offer hypotheses, use the extension word *if*, and capitalize on surprising results. Paul invited his students to create hypotheses to explain which numbers might be divisible by 3. Jacob used the important word *if* that enabled him to "pretend" that there were five seats in the car instead of four. *If* allows learners to go beyond the given and look at the present problem from a different perspective. David used Sharon's initial surprise at finding squares inside rectangles to set a new direction for her investigation. In these ways teachers can operationalize the mathematical thinking advocated by *Principles and Standards for School Mathematics* (National Council of Teachers of Mathematics [NCTM] 2000). This document uses a variety of verbs to describe the active role that learners must play in their mathematical learning: *explore, investigate, conjecture, predict, explain, discover, construct, describe*, and so on. All these verbs emphasize the importance of children making sense of mathematics for themselves. These verbs require children to be initiators, not receivers, of mathematical knowledge. Initiators pose problems, raise questions, offer hypotheses, and share their reasoning with peers. By encouraging the modification and extension of problem attributes, teachers can support children to be active constructors of their mathematical knowledge.

Problem posing is also connected to NCTM's Teaching and Assessment Principles. Assessment ought to be ongoing, helping teachers understand what children know and what next steps they need to plan for their students. By posing their own problems, children gain confidence in setting their own mathematical challenges as well as demonstrating mathematical competencies. For instance, Jacob's mother now knows that her son can add more than two addends. Jacob might pursue this idea further by adding other carloads of passengers. Paul's second graders demonstrated different theories of divisibility that they might explore further in relation to odd and even numbers. Sharon discovered an interesting mathematical pattern that she might extend to different-sized rectangles. Problem posing opens up new facets of a problem to explore and reveals to teachers new insights about their students' mathematical abilities.

Through problem posing, teachers can encourage learners to be mathematical "long-jumpers," not "high jumpers" (Domoney 1998). In the "high jump," the "bar" is set a specific height, such as the skill of adding two one-digit numbers. Some learners may be able to jump over this bar, and some may not. However, as teachers, we never really know how high they could have leaped. In contrast, the "long-jump" allows learners the opportunity to run and jump as far as they can. Problem posing represents this kind of open-ended mathematical event, giving teachers a more accurate and authentic assessment of what learners can do. Five-year-old Jacob showed that he could add multiples of 4 and 5, to his mother's amazement. The second graders demonstrated some understanding of the role of digits, and of odd and even numbers, in thinking about divisibility. Sharon showed a sense of the distributive property in the context of her square-number investigation. Long-jumpers not only display their mathematical talents but also gain confidence in themselves as mathematical thinkers.

The remainder of this article describes three additional strategies that teachers can use to promote a problem-posing environment. The first strategy involves children looking for patterns and developing generalizations. The second demonstrates how teachers can turn children's observations into mathematical extensions and variations. The third describes how teachers can capitalize on children's surprises, or unexpected results, to create further investigations.

TEACHERS INVITE A SEARCH FOR PATTERNS AND GENERALIZATIONS

Another way that teachers can promote a problem-posing stance is to invite children to look for patterns and seek generalizations. This is what occurred in Maria Sausa's fourth-grade class that was exploring patterns on a hundred chart. They were using different-shaped frames to examine certain sets of numbers. In this instance, they were using a frame composed of three squares in a row (fig. 11.1).

The children laid the frame down over different sets of three numbers on the chart and added to find the sum. They tried the following numbers.

$$1 + 2 + 3 = 6$$
$$7 + 8 + 9 = 24$$
$$12 + 13 + 14 = 39$$
$$79 + 80 + 81 = 240$$

1	2	3	4	5	6	7	8	9	10
11	12	13	14	15	16	17	18	19	20
21	22	23	24	25	26	27	28	29	30
31	32	33	34	35	36	37	38	39	40
41	42	43	44	45	46	47	48	49	50
51	52	53	54	55	56	57	58	59	60
61	62	63	64	65	66	67	68	69	70
71	72	73	74	75	76	77	78	79	80
81	82	83	84	85	86	87	88	89	90
91	92	93	94	95	96	97	98	99	100

Fig. 11.1

The teacher asked if the children noticed anything about the sum and its relationship to any of the addends. Several children saw that the sum was three times the middle number of the sequence. For instance, $2 \times 3 = 6$; $8 \times 3 = 24$; $13 \times 3 = 39$; $80 \times 3 = 240$. The teacher kept extending the children's thinking by asking, "What if you tried other numbers? Would you find the same relationship?" The children decided to test out numbers in the vertical columns, as shown below.

$$7 + 17 + 27 = 51$$
$$50 + 60 + 70 = 180$$
$$32 + 42 + 52 = 126$$

Here again they found that by tripling the middle number, they obtained the sum. Now they wondered about numbers on the diagonal. It is likely that the teacher's earlier support to try out different sets of numbers encouraged them to continue to go further. They tried again with the following:

$$9 + 18 + 27 = 54$$
$$24 + 35 + 46 = 105$$
$$77 + 88 + 99 = 264$$

Their conjecture was confirmed once again. The teacher then posed the question, "Will this work all the time?" She wanted them to go beyond these immediate examples and express the relationship in a more general way. Teachers who engage in problem posing not only push for problem variations but also challenge children to create generalizations from these patterns. While the children looked at their answers to explain what was happening, one child observed that the middle number was "like a balance because one of the other numbers is too big and the other too small." Another child saw that this difference from the middle number was the same. The children pursued this idea and noted that the horizontal numbers had a difference of 1, the vertical numbers had a difference of 10, and the diagonal numbers had a difference of 9 or 11, depending on the direction of the diagonal. There were several differences, but it was always the same difference from the middle number each time. The teacher led them into a discussion about the generalizability of their idea: No matter what the middle number n is, as long as the other numbers are the same difference a away from that number, the sum will be three times that middle number: $(n - a) + (n) + (n + a) = 3n$. Generalizations are an important labor-saving device of mathematicians, allowing them to set a general rule for conceptual understanding of a process or idea. Allowing children to pose these questions honors them as mathematical thinkers and problem solvers.

TURNING CHILDREN'S OBSERVATIONS INTO QUESTIONS

Good teachers often use children's observations as seeds for further problem-posing investigations. Extending and modifying a problem are two ways that teachers can promote this problem-posing strategy. Extending is a way for students to continue a particular pattern by finding other examples. Modifying a problem involves changing one or more of its attributes. The following examples demonstrate these aspects of the problem-posing process:

1. A kindergarten child shows you an ABABAB pattern she made out of red and yellow Unifix cubes. She says it's a pattern because it "goes back and forth."

   Extend: What if you kept adding more red and yellow cubes to your pattern?

   Modify: What if you tried using two other colors besides red and yellow?

2. A third grader remarks, "27 is an odd number because it has a 7."

   Extend: What if we looked at other numbers: 37, 47, 57? What do you notice?

   Modify: What about other numbers that have 7s in other places, such as 70 and 702?

3. A second-grade child says to a friend, "Hey, we're the same height, and we wear the same shoe size."

   Extend: What if we looked at other pairs of children to see if that is always true?

   Modify: What if we compared two other measurements, such as hat size and shoe size?

4. A fourth grader uses pattern blocks to make several large hexagons.

   Extend: How many different-sized hexagons can you make?

   Modify: What if you tried making larger versions of the other shapes?

5. A first grader holds up two fingers on each hand to show that "you can make four with two twos."

   Extend: What other ways can you show four with your fingers?

   Modify: What other numbers can you show with your fingers?

6. A fifth grader notices that the diagonal of a square cuts the area into two equal pieces.

   Extend: In what ways can we divide a square in half?

   Modify: In what ways can we divide other regular polygons in half?

As illustrated above, both extending and modifying involve posing new problems. However, extending is a strategy to continue the problem, such as finding more examples of odd numbers with a 7s digit. In contrast, modifying the problem involves changing at least one attribute, such as finding ways to divide other polygons in half. Children often make observations that can be extended or modified in these different directions. Teachers build a supportive and intellectually stimulating learning community when they recognize the observations of their students and then challenge them to investigate their ideas further.

CAPITALIZING ON UNEXPECTED RESULTS

Problem posing often arises when children are dissatisfied, or surprised, with a particular situation and naturally suggest revisions. This is in fact what happened in Susan Cvitkovich's fifth-grade class. The children had been introduced to a trading game ("race for a flat") using base-ten blocks. They rolled one die and kept accumulating on their board the number of units rolled, and then eventually traded units for longs, and then ten longs for a flat. The children enjoyed the game but were surprised that they couldn't get to a flat more quickly. One child suggested that they play by a different rule: "What if we multiply the number we roll by itself?" So if a child rolls a 5, she would calculate 5×5 and place the value of 25 units on the board. The teacher encouraged the children to test out this variation. The children were amazed at how quickly the game proceeded. After a few games using this new rule, another child suggested, "What if you triple the number you roll?" predicting that the class would complete the game even more quickly. However, after playing one game with this new rule, another child remarked, "I'm really surprised. This way is slower than when you multiply a number by itself." It was this surprise that led them to investigate the situation even further. The teacher invited the children to create a chart (fig. 11.2) so that they could compare the results of these game variations. (Although the greatest possible value on a die is 6, the children wanted to continue the chart to see more of the patterns.)

Number	Tripled	Times itself (squared)
1	3	1
2	6	4
3	9	9
4	12	16
5	15	25
6	18	36
7	21	49
8	24	64
9	27	81

Fig. 11.2

The children noticed that only the numbers 1 and 2 produced tripled answers that were greater than their comparable squared answers, and that 3 was the only number that produced the same answer for tripling and squaring. In analyzing this finding, the children realized that tripling the number 3 was really the same as squaring it. They also observed that the tripled column increased by 3s and the squared column increased by consecutive odd numbers. They saw that the difference between squared and tripled numbers had patterns (fig. 11.3). For example, starting in row 5, 25 – 15 = 10, 36 – 18 = 18, 49 – 21 = 28, 64 – 24 = 40, and 81 – 27 = 54. They also noticed that the differences between the underlined answers were consecutive even numbers: 8, 10, 12, and 14. It was clear that squaring a number generated a larger increase than tripling that number.

Number	Tripled	Differences		Times itself (squared)
1	3			1
2	6			4
3	9	0		9
			4	
4	12	4		16
			6	
5	15	10		25
			8	
6	18	18		36
			10	
7	21	28		49
			12	
8	24	40		64
			14	
9	27	54		81

Fig. 11.3

As the children talked further about their surprise, the teacher realized that some of the children were confusing squaring a number with doubling a number. She invited the children to make another chart so that they could compare all three variations (see fig. 11.4).

The children noticed new relationships with this revised chart: The doubling column increased by 2s. The difference between the numbers in the doubled and squared columns (beginning with the number 3) was 3, 8, 15,

Number	Doubled	Tripled	Squared
1	2	3	1
2	4	6	4
3	6	9	9
4	8	12	16
5	10	15	25
6	12	18	36
7	14	21	49
8	16	24	64
9	18	27	81
10	20	30	100

Fig. 11.4

24, 35, 48, which are numbers that are 1 less than a square number, and they too increase by consecutive odd numbers. Other observations included the following:

- The difference between the doubled and tripled columns increased by 1 each time.
- Two was the only number that had the same answer for doubling and squaring itself. The children compared this result to what they had discovered about 3 in the previous chart.
- Tripling a number always produced an answer greater than doubling a number.
- Only the numbers 1 and 2 produced doubled answers that were greater than or equal to their squared answers.

The class ended their investigation at this point, but from a problem-posing perspective, the work could have continued. For instance, we might think about performing other operations on numbers and comparing the results: How does squaring a number compare with cubing a number? How does cubing a number compare with tripling a number? How many times do you have to double a number until you reach or pass its square? Knowing that there is always unfinished business to explore is a healthy attitude to have as learners. It keeps us humble and gives us more possibilities to mull over.

This last mathematical story demonstrates how valuing surprise contributes to a healthy and productive classroom environment. By valuing surprise, teachers legitimize unexpected outcomes and support a risk-taking stance (Whitin and Whitin 2000). Teachers also send the implicit message that children who are surprised are respected sense makers who want to know why things did not turn out as expected.

PROMOTING PROBLEM POSING

Teachers who employ the strategy of problem posing grant many benefits to their students. Problem posing promotes many of the attitudinal goals of the NCTM *Standards* by encouraging risk taking, perseverance, curiosity, skepticism, and the postponement of judgment. It encourages learners to welcome puzzlement, expect the unexpected, and have faith in their sense-making efforts. It grants children a sense of ownership and responsibility for the exploration. It involves the hunt for patterns. It also sets children out on trails that are not well marked, and this is a kind of traveling that mathematicians relish.

Teachers can support a problem-posing stance in their classroom by keeping some of these suggestions in mind:

1. Encourage children to be careful observers. Ask questions such as "What do you notice?" and "What do you find interesting?" Noticing enables the class to see the attributes of a problem or situation. Once we know the attributes, we can then begin to extend or modify them to create related problems.
2. Encourage children to go beyond the given problem by using the word *if*. This small word can take us on long journeys: What if we looked for other squares inside larger rectangles? What if we looked at other sets of three numbers on our hundred chart?
3. Think about extensions and modifications yourself in all the mathematical work you do with children. Even a familiar mathematical problem or task can become more interesting through a slight change in its focus. In this way teachers and students alike can feel the sustaining power of a mathematically inquisitive mind.

Teachers who have worked to establish a problem-posing environment in their classroom have testified to its power:

- I realized that I wasn't talking as a teacher looking for the right answer, but I was talking as a fellow learner.
- Problem posing taught me that a problem is not as simple as the question that asked it.
- Problem posing helps the teacher be a risk taker, fear breaker, and freedom maker.

Students, too, have recognized the benefits of a problem-posing stance and what that means for their own attitudes and dispositions toward mathematics:

- I discovered that I had a great imagination [grade 5].
- When I changed the rules, I became the teacher and not the student [grade 5].

- I like it [problem posing] because I like a challenge. I feel like a detective, and I'm a good detective because I found clues [grade 7].

These reflections show how problem posing can break down the dichotomy between teacher and learner, building an environment that celebrates the thinking and imaginative ideas of all learners.

REFERENCES

Brown, Stephen, and Marion Walter. *The Art of Problem Posing.* Hillsdale, N.J.: Lawrence Erlbaum Associates, 1983. Reprint, 1990.

Domoney, Bill. "Numeracy Teaching in the Primary Classroom." Paper presented at the European Council of International Schools Conference, Hamburg, Germany, November 1998.

National Council of Teachers of Mathematics (NCTM). *Principles and Standards for School Mathematics.* Reston, Va.: NCTM, 2000.

Whitin, Phyllis, and David J. Whitin. *Math Is Language Too: Talking and Writing in the Mathematics Classroom.* Urbana, Ill.: National Council of Teachers of English; and Reston, Va.: National Council of Teachers of Mathematics, 2000.

12

Promoting Equity in Mathematics Education through Effective Culturally Responsive Teaching

Joan Cohen Jones

Effective teaching is an essential component of equity in mathematics education.

> Excellence in mathematics education requires equity—high expectations and strong support for all students.... All students, regardless of their personal characteristics, backgrounds, or physical challenges, must have opportunities to study—and support to learn—mathematics. (National Council of Teachers of Mathematics [NCTM] 2000, p. 12)

This article takes a closer look at the characteristics of effective culturally responsive teachers and the classrooms they create; that is, those teachers who, through their behaviors and expectations, create an equitable learning environment for all their students. It should be noted that the characteristics of effective culturally responsive teachers are compatible with the characteristics of any successful teacher and are supported by extensive research on teacher effectiveness (Irvine 2001).

Characteristics of Effective Culturally Responsive Teachers

Considerable research has been conducted to identify the characteristics of effective culturally responsive teachers. Researchers have observed specific teachers' behaviors and generalized from their observations.

Ladson-Billings (1995) studied eight effective teachers of African American students, one of whom was Ms. Rossi, an Italian American woman in her forties (pp. 134–35):

> From a pedagogical standpoint, I saw Ms. Rossi make a point of getting every student involved in the mathematics lesson. She continually assured students that they were capable of mastering the problems. They cheered each other on and

> celebrated when they were able to explain how they arrived at their solutions.... Ms. Rossi moved around the room as students posed questions and suggested solutions. She often asked, "How do you know?" to push the students' thinking.

Ladson-Billings concluded that Ms. Rossi treated all her students as competent and capable of learning, used their prior knowledge to help then build an understanding of new concepts, and engaged them in meaningful learning during the entire class period.

Gutstein et al. (1997) worked with eight teachers at the Diego Rivera Elementary and Middle School, where 96.4 percent of the students were Mexican American and 99 percent of students were low income. Gutstein observed teachers, helped plan lessons, team-taught, and worked individually with children. Ms. Herrera, a bilingual third-grade teacher at the school, illustrated another example of effective teaching:

> We have repeatedly seen Ms. Herrera disagree with her students, make them justify their responses, ask for alternative explanations, screw up her face in disbelief, and challenge their thinking.... Ms. Herrera says that she sees a direct relationship between "pushing them to say why is a third bigger than an eighth" and "helping them become leaders." (Gutstein et al. 1997, p. 720)

The authors concluded that effective culturally responsive teachers like Ms. Herrera teach mathematics with a broader framework in mind. That is, their mathematics teaching prepares students to become critical thinkers, stand up for what they believe, and be active participants in society.

Dance, Wingfield, and Davidson (2000) studied a well-respected teacher at a predominantly African American inner-city "school within a school" with a record of "student persistence in the study of mathematics through high school and beyond" (p. 35). Two factors seemed to promote mathematics achievement: "a sense of community" and an "atmosphere of challenge" (p. 35).

The sense of community was fostered by mutual respect, the encouragement of individuality, and a sense of safety. The teacher did not control the discussion but let it flow naturally. When the teacher made an error, she openly acknowledged her mistakes and sincerely thanked her students for bringing the error to her attention.

The atmosphere of challenge was perceived as a necessary partner to the sense of community. The term *challenge,* used in this context, described both the types of problems presented and the teacher's actions. Mathematics problems that were both difficult and interesting were deemed challenging. To qualify as interesting, the problems had real-world contexts or were connected to other school subjects. The teacher contributed to an atmosphere of challenge by her "interest in the problems, her interest in multiple solutions, and her demonstrated respect for students' solutions different from her own" (Dance, Wingfield, and Davidson 2000, p. 41).

Several characteristics of culturally responsive teachers are listed below in two categories: pedagogy and beliefs. Pedagogy includes instructional behaviors that foster effective culturally responsive teaching. Beliefs include beliefs and attitudes that support these instructional behaviors. Each characteristic is then discussed in greater detail, followed by a discussion of the classroom atmosphere that supports culturally responsive teaching.

Pedagogy includes

- a knowledge of subject matter;
- an ability to listen to and question students to learn about their thinking;
- a willingness to use cultural knowledge to make connections to new knowledge.

Beliefs include

- an understanding of, and respect for, students' cultural beliefs and values;
- a respect for students' ability and competence;
- an ability to be reflective.

Pedagogy

Teachers who are proficient in subject-matter knowledge tend to be more flexible in their choices of content, pedagogy, and assessment (Grant 1995). The more we know, the more flexible we tend to be in how we present material, what questions we ask, how much authority we share, and how we assess knowledge.

Teachers who listen carefully to their students learn what their students know, what they misunderstand, what is important for them to learn, and productive ways for them to learn. Listening skills are interrelated with subject-matter knowledge, confidence, and efficacy. Teachers who are knowledgeable and confident can direct attention away from themselves without losing the momentum of the lesson.

Good questions reinforce and extend students' learning, encourage students to give explanations, and inspire students to rely on themselves and their classmates, building self-reliance, cooperation, respect, and confidence (Dougherty, Slovin, and Matsumoto 1999). Examples of effective questions include the following:

- How do you know?
- Will that always work?
- Have we seen something like this before?
- Can you solve this another way?

White (2000) worked with a first-year third-grade teacher, Ms. Tyler, as part of Project IMPACT (Increasing the Mathematical Power of All Children and Teachers). Prior to her work with Project IMPACT, Ms. Tyler seemed to follow the recommendations of the NCTM *Standards* by using hands-on materials, group learning, children's literature, and learning centers to teach mathematics. However, her questioning skills needed improvement. During initial observations, Ms. Tyler did not ask students to explain correct answers and sometimes misinterpreted students' responses. When a student answered incorrectly, Ms. Tyler "led him to the correct answer through her questioning" (p. 23). When a student answered correctly, Ms. Tyler assumed that the student's reasoning was also correct. As part of Project IMPACT, Ms. Tyler attended a twenty-two-day summer in-service program and also received on-site support from a mathematics specialist during the next academic year. With the help of Project IMPACT, Ms. Tyler learned to ask students questions that focused on their thinking and how they arrived at their answers. She asked students to explain their answers and encouraged student discourse by asking them if they agreed with other students' explanations and answers.

Effective teachers increase wait time when asking or answering questions. That is, when listening to students' questions, they pause several seconds before trying to answer the question themselves or asking someone else to answer it. When asking questions of students, they wait several seconds longer than usual for an answer. This can be especially beneficial to students from ethnically or culturally diverse backgrounds, who may take longer to answer.

Effective culturally responsive teachers connect cultural knowledge to new knowledge in many exciting ways. They use "authentic cultural data, literature, music, art, artifacts, ... and cultural history" (Armento 2001, p. 27) in their curriculum to represent students' diverse backgrounds.

As one example, children's literature is an excellent vehicle to bridge the gap from cultural knowledge to mathematics learning. *A Grain of Rice* (Pittman 1986) teaches about the culture of ancient China while examining the properties of 2^n. In the story a peasant receives one grain of rice, doubled each day for 100 days, as a reward for saving the life of the emperor's daughter. As students compute the amount of rice rewarded each day and the total grains of rice rewarded from the first day, they investigate interesting mathematical patterns (Thompson, Chappell, and Austin 1999). The story can be modeled in class by using counters, candy, or beans to represent the rice. As children double the amount of "rice" rewarded each day and count the amount of rice accumulated, they observe how quickly the total reward increases and get hands-on experience with exponential functions.

As another example, the textiles produced by a culture serve as an effective means of learning about the culture. The informal mathematics used to design and produce textiles can stimulate the learning of traditional mathematics. Although the lesson ideas discussed below use Navajo rugs, other textile choices might include Kente cloth (Jones 2001), Hmong paj ntaub (Jones 1999), and Mexican embroidery.

Navajo rugs often illustrate bilateral symmetry and use polygons in their design (Jones 2002). As students examine pictures of Navajo rugs, ask them to identify the polygons woven into the designs and the overall symmetry of the rugs. Challenge students to design and color their own rug patterns on sheets of 10 × 10 grid paper. Instruct them to use the entire sheet of grid paper for their design and to color their design in three different colors, using markers or crayons. Ask students to calculate the area of each color used, with each grid square representing one square unit of area. This gives students the opportunity to investigate the areas of regular and irregular regions. To extend the activity, ask students to express the area of each color used in their design as a fraction of the area of the entire rug.

Beliefs

Culturally responsive teachers respect students' cultural differences and seek to understand their students' cultural backgrounds. Although it is not always possible to have expert knowledge of every culture, especially in a multiracial classroom, effective multicultural teachers seek to expand their knowledge of their students' cultural beliefs, values, attitudes, and patterns of interaction. Culturally responsive teachers get to know their students well, so that they can forge close relationships with their students and form strong ties with the community in which their school is located.

Effective culturally responsive teachers hold high expectations for all their students. They respect their students' ability and competence by giving them high-level intellectual tasks that require complex processing and critical thinking. Ms. Rossi demonstrated this by questioning her students and turning questions asked of her into questions for her students, thus helping her students understand that she saw them as knowledgeable and responsible learners.

Culturally responsive teachers need to be reflective practitioners (Irvine 2001). That is, they need to monitor, review, and revise their practices, instructional choices, and methodology on a regular basis to ensure that they are being fair and open-minded, offering high-level tasks for their students, and using their students' cultural backgrounds to connect to new concepts. An important step in reflection is to think about one's own cultural heritage, to understand one's own biases, and to realize that education is often laden with the values of the dominant culture.

CLASSROOM ATMOSPHERE

Effective culturally responsive teachers create classroom atmospheres that provide equitable learning environments for all students.

> Students are treated as individuals, are provided with equal access to learning resources, and are nurtured as special and creative young people ... educators and students discuss the issues and problems that arise and make consensus decisions about how these matters will be handled. (Armento 2001, p. 22)

Effective culturally responsive teachers may have very different teaching styles, but their classrooms have a number of similarities. They are structured and orderly. Expectations for students are clearly stated and understood. Although the teacher is definitely in charge, there is a sense of interdependence. Students and teacher work together and share responsibility for learning. Teachers model compassion, consideration, and respect for students in a manner consistent with their own personalities.

Classrooms that facilitate students' learning share a number of common characteristics. Three important ones are discussed below:

- Students' choice
- Cooperative learning
- The classroom as a community

Students' Choice

The effective multicultural classroom offers students choices in their assignments, with whom they work, how they respond, and how they are assessed. Students' choice is especially important for students from ethnically or culturally diverse backgrounds, who may have different learning styles from their peers from the dominant culture. When developing classroom activities, effective teachers offer choices in assignments and types of assessments; give students the opportunity to work independently, in pairs, or in small groups; and allow students to respond in oral or written form, individually, in teams, or as a class. Choices such as these allow each student to participate in the manner that is most comfortable for him or her, both academically and culturally. Thus, each student has the opportunity to use his or her strengths in learning new concepts and demonstrating what he or she has learned. Student choice leads to success and self-esteem. It helps build a sense of responsibility for one's own learning as well as responsibility to one's peers. Student choice also leads to teacher choice, allowing teachers to be more creative in their choice of content, presentation style, and examples.

Although many teachers believe the opposite, student choice actually helps in classroom management. I have been in many classrooms where a student, who would have been happy to write a written report, is forced to present an oral report with negative results. I have seen students refuse to go to the chalkboard alone, although they would willingly have gone with a partner. Conversely, I have also been in classrooms where choice defused a potentially difficult situation and allowed class time to be spent on what is really important—learning. When students are given choices, the classroom becomes more interesting, comfortable, spontaneous, and meaningful for everyone.

Cooperative Learning

Effective culturally responsive teachers frequently use cooperative groups in their mathematics classrooms. Research has shown that students from ethnically or culturally diverse backgrounds learn more effectively in cooperative learning groups and that cooperative groups help to increase self-esteem for these learners (Sleeter 1997).

Malloy (1997) determined that African American students perform better in cooperative learning groups because of the traditions of their culture:

> Harmony within the community and a holistic perspective provide a basis for social interaction, which seems to be crucial in African American culture. The African American community focuses more on people and their activities than on things, and the interdependence of people and their environment is respected and encouraged.... One instructional application of interdependence is cooperative learning. (P. 24)

African American students tend to work well when decisions, responsibilities, and materials are shared among the group.

Cooperative group learning is also beneficial for language-minority students, who may not perform well in whole-group mathematics instruction. Competitive situations may not be compatible with their cultural backgrounds, and they may not feel comfortable asking questions in the whole-class environment because of perceived or real difficulties with English (Sleeter 1997). These students are simultaneously trying to learn mathematics and communicate in a new language. The small group provides a safe place for them to work with others, ask questions, express themselves in English, and learn new concepts.

Effective culturally responsive teachers are aware of the learning styles described above but respond to each student as an individual learner with unique experiences and needs.

> One must be careful not to overgeneralize when making statements ... about the dominant patterns of groups. It must always be remembered that individuals within any group may differ dramatically from these dominant patterns and that we cannot know about individuals, their beliefs, their behavior, and their preferences simply because of their membership in certain groups. (Armento 2001, p. 29)

The Classroom as a Community

Effective culturally responsive teachers create communities within their classrooms. Their classrooms are safe havens, places where each person feels cared about and cares about others. They have traditions, modes of interaction, and shared history. Students find reassurance and support in their classrooms. In Ms. Rossi's (Ladson-Billings 1995, 1997) class, students cheered each other on as they solved difficult mathematics problems. In Ms. Herrera's class (Gutstein et al. 1997), the teacher fostered a democratic classroom atmosphere where all students felt comfortable to share their own methods for solving problems. Students were encouraged to speak out and become leaders. At Drew High School (Dance, Wingfield, and Davidson 2000), the teacher fostered a sense of community by participating actively in class discussions without dominating them, building a sense of mutual respect and safety in the class by admitting her own errors and correcting them, and respecting individual differences.

Although the classroom culture affects all students, it seems to be especially important to students from ethnically or culturally diverse backgrounds. In fact, these students have reported that a personal relationship with their teacher played a decisive role in their performance in mathematics. Students who have good personal relationships with their teachers feel more confident in class, are more engaged in learning, and are more likely to take additional mathematics classes (Walker and McCoy 1997).

The effective culturally responsive teacher and his or her students create a very special place where students and teacher feel free to share personal stories about themselves and their families. A common bond is created. Teachers show their support with frequent words of praise and encouragement. Teachers understand the special role they play in their students' lives.

CONCLUDING THOUGHTS

It is important to identify the characteristics of effective culturally responsive teachers and classrooms. But classrooms are in schools, and schools are part of communities. Culturally responsive teachers are more effective if they have strong ties with the community, communicate with parents, and understand and respect the cultural backgrounds of their students and their families. Teachers are more effective if administrators, the school, and the school system have the same goals and provide support for teachers.

REFERENCES

Armento, Beverly. "Principles of a Culturally Responsive Curriculum." In *Culturally Responsive Lessons for the Elementary and Middle Grades: A Guide for Preservice Teachers,* edited by Beverly Armento and Jacqueline Jordan Irvine, pp. 19–33. New York: McGraw-Hill, 2001.

Dance, Rosalie A., Karen H. Wingfield, and Neil Davidson. "A High Level of Challenge in a Collaborative Setting: Enhancing the Chance of Success for African American Students in Mathematics." In *Changing the Faces of Mathematics: Perspectives on African Americans,* edited by Marilyn E. Strutchens, Martin L. Johnson, and William F. Tate, pp. 33–49. Reston, Va.: National Council of Teachers of Mathematics, 2000.

Dougherty, Barbara J., Hannah Slovin, and Annette N. Matsumoto. "Creating a Classroom Culture: A Pacific Perspective." In *Changing the Faces of Mathematics: Perspectives on Asian Americans and Pacific Islanders,* edited by Carol A. Edwards, pp. 55–64. Reston, Va.: National Council of Teachers of Mathematics, 1999.

Grant, Carl A. "Critical Knowledge, Skills, and Experiences for the Instruction of Culturally Diverse Students: A Perspective for the Preparation of Preservice Teachers." Paper presented at the Invitational Conference on Defining the Knowledge Base for Urban Teacher Education, Emory University, Atlanta, November 1995.

Gutstein, Eric, Pauline Lipman, Patricia Hernandez, and Rebeca de los Reyes. "Culturally Relevant Mathematics Teaching in a Mexican American Context." *Journal for Research in Mathematics Education* 28 (December 1997): 709–37.

Irvine, Jacqueline J. "The Critical Elements of Culturally Responsive Pedagogy: A Synthesis of the Research." In *Culturally Responsive Lessons for the Elementary and Middle Grades: A Guide for Preservice Teachers,* edited by Beverly Armento and Jacqueline Jordan Irvine, pp. 2–17. New York: McGraw-Hill, 2001.

Jones, Joan C. "Hmong Needlework and Mathematics." In *Changing the Faces of Mathematics: Perspectives on Asian Americans and Pacific Islanders,* edited by Carol A. Edwards, pp. 13–20. Reston, Va.: National Council of Teachers of Mathematics, 1999.

———. "Craft Patterns and Geometry." In *Culturally Responsive Lessons for the Elementary and Middle Grades: A Guide for Preservice Teachers,* edited by Beverly Armento and Jacqueline Jordan Irvine, pp. 54–101. New York: McGraw-Hill, 2001.

———. "Finding Cultural Connections for Teachers and Students: The Mathematics of Navajo Rugs." In *Changing the Faces of Mathematics: Perspectives on Indigenous People of North America,* edited by Judith E. Hankes and Gerald R. Fast, pp. 175–81. Reston, Va.: National Council of Teachers of Mathematics, 2002.

Ladson-Billings, Gloria. "Making Mathematics Meaningful in Multicultural Contexts." In *New Directions for Equity in Mathematics Education,* edited by Walter G. Secada, Elizabeth Fennema, and Lisa B. Adajian, pp. 126–45. New York: Cambridge University Press, 1995.

———. "It Doesn't Add Up: African American Students' Mathematics Achievement." *Journal for Research in Mathematics Education* 28(December 1997): 697–708.

Malloy, Carol E. "Including African American Students in the Mathematics Community." In *Multicultural and Gender Equity in the Mathematics Classroom: The Gift of Diversity,* 1997 Yearbook of the National Council of Teachers of Mathematics (NCTM), edited by Janet Trentacosta, pp. 23–33. Reston, Va.: NCTM, 1997.

National Council of Teachers of Mathematics (NCTM). *Principles and Standards for School Mathematics.* Reston, Va.: NCTM, 2000.

Pittman, Helen C. *A Grain of Rice.* New York: Bantam Skylark, 1986.

Sleeter, Christine E.. "Mathematics, Multicultural Education, and Professional Development." *Journal for Research in Mathematics Education* 28 (December 1997): 680–96.

Thompson, Denisse R., Michaele F. Chappell, and Richard A. Austin. "Exploring Mathematics through Asian Folktales." In *Changing the Faces of Mathematics: Perspectives on Asian Americans and Pacific Islanders,* edited by Carol A. Edwards, pp. 1–11. Reston, Va.: National Council of Teachers of Mathematics, 1999.

Walker, Erica N., and Lea P. McCoy. "Students' Voices: African Americans and Mathematics." In *Multicultural and Gender Equity in the Mathematics Classroom: The Gift of Diversity,* 1997 Yearbook of the National Council of Teachers of Mathematics (NCTM), edited by Janet Trentecosta, pp. 71–80. Reston, Va.: NCTM, 1997.

White, Dorothy Y. "Reaching All Students Mathematically through Questioning." In *Changing the Faces of Mathematics: Perspectives on African Americans,* edited by Marilyn E. Strutchens, Martin L. Johnson, and William F. Tate, pp. 21–32. Reston, Va.: National Council of Teachers of Mathematics, 2000.

13

Enhancing the Mathematical Understanding of Prospective Teachers
Using Standards-Based Grades K–12 Activities

Charlene E. Beckmann

Pamela J. Wells

John Gabrosek

Esther M. H. Billings

Edward F. Aboufadel

Phyllis Curtiss

William Dickinson

David Austin

Alverna Champion

Recent reports from national agencies and professional organizations raise several concerns about the mathematical preparation of teachers. These reports indicate that prospective teachers need mathematics courses that develop a deep understanding of the mathematics they will teach. What often happens, unfortunately, is that the more mathematics prospective teachers take, the further away they move from the curriculum they will ultimately teach, better preparing them for graduate school than for the classroom (Usiskin 2001). Too often, courses do not make evident how undergraduate mathematics applies to teaching grades K–12 mathematics.

Many reports (CBMS 2001; NCMST 2000; NCTM 2000) recommend that universities create a variety of mathematics courses for grades K–12 teachers separate from those offered for other clients such as engineers. However, most mathematics departments are not large enough to offer separate courses for each constituent. As a result, mathematics departments face the dilemma of making mathematics meaningful and applicable for prospective teachers while still maintaining a high level of mathematical content for all clients. This article describes work being done at Grand Valley State University (GVSU), a comprehensive Midwestern university, to address these serious concerns.

AN OVERVIEW OF THE PROJECT AND GOALS

Supported by grants from the Pew Faculty Teaching and Learning Center at GVSU, mathematics, mathematics education, and statistics faculty members have joined together to enhance six core courses in the mathematics major: Communicating in Mathematics (an introduction to proof course), Euclidean Geometry, Probability and Statistics, Linear Algebra, Modern Algebra, and Discrete Mathematics. All mathematics majors, including both prospective elementary and secondary school teachers, take these courses. Exemplary grades K–12 mathematics activities are used to introduce important concepts in these core courses. These activities provide prospective teachers an access point to the complex mathematics of these college courses, helping them deepen their understanding of mathematics and experience how these concepts might be learned in grades K–12 classrooms. The inclusion of *Standards*-based grades K–12 materials reminds all majors of where they have encountered core concepts in previous courses while they experience the advanced mathematical and statistical understanding required of grades K–12 students. Consequently, prospective teachers are confronted with the need to make connections among important concepts and to understand them at a college level in order to teach these concepts to children.

The use of grades K–12 activities enriches the overall quality of these core courses by promoting active learning and oral communication through classroom discourse. Written communication occurs through carefully chosen or designed assessment items. Students' interest levels increase as the use of motivating materials promotes a desire for further study of related mathematical and statistical concepts.

To achieve the project's goals, teams of one mathematics educator and at least one mathematician or statistician study how the content of each core course aligns with current national standards and recommendations (e.g., NCTM 2000; CBMS 2001). Then these teams determine which topics can be enhanced through the use of exemplary grades K–12 mathematics materials. Finally, the teams find, adapt, or create grades K–12 level materials to use in college-level mathematics courses. The teams have used a variety of sources, including activities from grades K–12 mathematics curriculum projects supported by the National Science Foundation (NSF) and journal articles, especially those found in the National Council of Teachers of Mathematics (NCTM) journals.

ACQUIRING DEEP AND CONNECTED MATHEMATICAL KNOWLEDGE

A unifying goal of this project is to deepen all students' mathematical content knowledge. Understanding deepens as students establish connections between conceptual and procedural knowledge (Hiebert and Carpenter 1992). In the six core courses, many different mathematical and statistical concepts and ideas are explored.

In this article, one particular aspect of mathematical knowledge, reasoning and proof, is examined. In all the core courses, students are expected to conjecture, reason, and communicate logically and to write valid proofs. Prospective teachers must also understand the need for proof and grow in their ability to analyze the reasoning of others. Since conjecture and proof are fundamental, many materials were created around this theme and embedded in the core courses.

The following sections discuss three activities developed to promote conjecturing, reasoning, and proof. First, in a problem involving probability, students reason informally by making and testing conjectures. Then, in an activity involving patterning and recursion, students apply informal reasoning and are introduced to a way of formalizing their arguments. Finally, in a geometry activity, formal proof methods are applied.

MAKING AND TESTING CONJECTURES IN A PROBABILITY AND STATISTICS COURSE

Many students have difficulty making statistical inferences and interpreting probability statements (Garfield and Ahlgren 1988). Consequently, a variety of tasks were designed that address common statistical and probabilistic misconceptions. The tasks explore important concepts in the course, including sampling, graphical and numerical summaries of data, basic probability concepts including combinations and permutations, probability distributions of random variables, sampling distributions, confidence intervals, and hypothesis testing. Most of the activities require students to discover important concepts before formal discussion.

The Mighty Thumbtack activity (fig. 13.1), versions of which are found in a number of elementary grades textbooks, builds on students' prior knowledge of frequency tables, relative frequency, and formal graphing. Students make and test conjectures about how a dropped thumbtack lands. The activity begins by asking students to make a reasoned conjecture about whether the probability of a dropped thumbtack landing point up is less than, greater than, or equal to 0.5.

Students' writing about their predictions reveals common misconceptions of statistical terminology and concepts. For example, "... statistically there is a 50/50 chance to get up or down. But that is unlikely because of certain factors. It depends on how the tack is dropped and how far from the table. Each time you have to do the trials the same way to get an accurate result." At first the student commits the common fallacy in probabilistic reasoning of thinking that all outcomes of experiments are necessarily equally likely. But she

The Mighty Thumbtack

Is a dropped thumbtack more likely to come to rest point up or point down (i.e., not point up)? Or are the chances equal? Write your guess (up, down, or equal) on the line below.

Guess ________________

1. Write a short paragraph explaining why you guessed as you did.
2. Write a short paragraph describing how you would investigate the thumbtack question. Think and write about issues related to independence of trials and repeatability of your experiment.
3. With the thumbtack that you have been given, carry out the following investigation: Drop the thumbtack from a height of 1 foot 20 times. Keep track of the results of your experiment and display those results in a table.
4. Make a line graph with the **cumulative proportion of point up** on the vertical axis and the **trial number** on the horizontal axis.
5. Describe what you see in your line graph.
6. Compare your line graph to that of the other students in your group. Discuss similarities and differences.
7. Discuss your proposed investigation methods in (2) with one other person in your group. Compare and contrast your two methods. What are the strengths and weaknesses of your method and the method of the person sitting next to you?

Fig. 13.1. The Mighty Thumbtack activity for the Probability and Statistics course

senses that 50/50 is not correct and dismisses that conjecture. Additionally, she fails to make a distinction between the meaning of probability as the likelihood of an event's occurrence and the odds in comparing two events. Her response, "Each time you have to do the trials the same way to get an accurate result," indicates that she has an intuitive understanding of the importance of repeated trials under the same experimental conditions. This low level of statistical knowledge is typical of students at the beginning of an introductory course.

Students test their thumbtack conjectures by dropping a thumbtack 20 times from a height of one foot and recording how often the thumbtack comes to rest point up. Students tabulate the cumulative number of times the thumbtack lands point up and make a line graph of the cumulative proportion of points up against the trial number. Students compare line graphs in small groups and reason that data from 20 trials are highly variable from student to student. However, when the entire class's data are plotted, the successive proportions of cumulative points up begin to converge to a value (≈ 0.40) called the probability of landing point up. Students see that the two outcomes are not equally likely. In addition, they discover a fundamental principle, the Law of Large Numbers. Later, the instructor references the activity when formally stating the Law of Large Numbers: As the number of trials increases, the relative frequency of an event's occurrence approaches the theoretical probability of the event.

The Mighty Thumbtack activity and other investigative activities lay the foundation for the course. They benefit students in multiple ways. First, they increase students' content knowledge by helping them confront and correct misconceptions. Students build on the knowledge gained from this activity to collect and use data to test future conjectures. Second, this activity provides a concrete reference point to which students and instructor can refer throughout the semester, as students explore concepts such as experimental design, probability, sampling distributions, and statistical inference. By integrating concepts through unifying activities, prospective teachers see that probability and statistics is a body of connected knowledge and not simply a collection of formulas and tricks. They also realize that topics such as making and testing conjectures through data collection and the Law of Large Numbers are explored as early as elementary school.

Developing an Inductive Argument in a Proof-Writing Course

In many mathematics programs, there is a first proof course that serves as a bridge between the calculus sequence and upper-level courses. GVSU's Communicating in Mathematics (CIM) course serves this purpose. In this course, students are expected not only to make and test conjectures, but also to learn formal proof methods while they develop mathematical reasoning skills and deepen their understanding of logic, set theory, number theory, and functions.

The CIM team developed more than a dozen activities on topics such as functions, greatest common divisor, conditional statements, set theory, and mathematical induction. The Staircase Problem (see fig. 13.2), a middle-grades patterning activity modified from Schoenfeld et al. (1999), helps students use inductive reasoning (generalizing from examples) in order to develop an informal understanding of mathematical induction (a form of deductive reasoning in which a proof is built through proving the beginning situation and then proving that given any instance in a sequence one can build the next instance in the sequence).

In the CIM course, the Staircase Problem provides a concrete reference on which to link students' future work with mathematical induction. This activ-

The Staircase Problem

The staircases shown below are made with toothpicks.

1. Look for a pattern in the number of toothpicks in the *perimeter* of each staircase.
2. Use the pattern from (1) to determine the perimeter of Staircase 5.

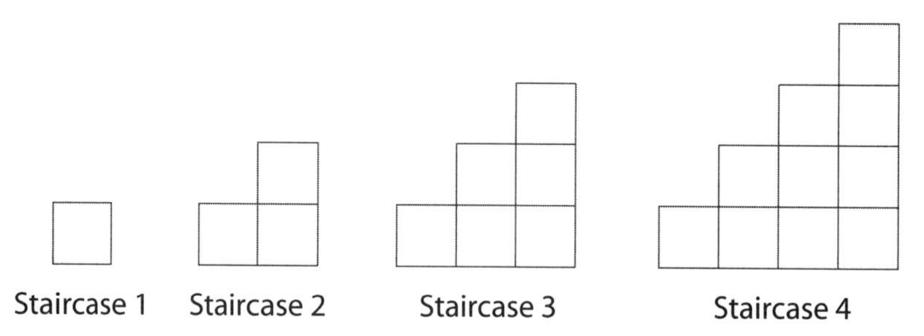

3. How can you use the perimeter of Staircase n to find the perimeter of the next staircase (Staircase $n + 1$)?
4. Write an argument to convince a friend who is doubtful that your answer to (3) is correct. Use the pictures of the staircases given above to show your friend how the perimeter increases going from Staircase n to Staircase $n + 1$.

Fig. 13.2. The Staircase Problem for the Communications in Mathematics course

ity allows students to reason inductively as they analyze and extend patterns in order to find the number of toothpicks needed for the perimeter of any staircase and to argue informally that the pattern they see in the first four staircases will continue for all staircases. In addition, the instructor then uses this activity to extend students' reasoning to create a proof by mathematical induction and to promote the need for this type of proof in general.

This activity has been used several times. The following discussion provides a representative range of responses that occur when students explore this activity. Groups present their patterns and conjectures to their classmates. Most of these groups use inductive reasoning to answer the third and fourth questions. They clearly extend what they know about the geometry of the perimeter of one staircase to determine the perimeter of the next staircase. For example, in one CIM class, three groups saw that one row (or column) of squares needed to be added to the previous staircase to make the next staircase. As shown in figure 13.3a, the groups that took this approach showed how adding a new row (which is not shaded in the diagram) to a staircase caused the perimeter of the staircase to increase by four toothpicks. Three new toothpicks are needed to complete the extra square in the new row and one new toothpick is needed to close off the row (slash marks indicate a new toothpick added to the perimeter).

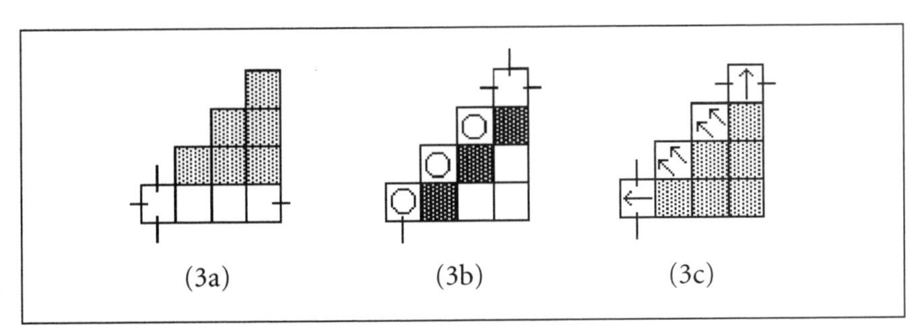

Fig. 13.3. Diagrams showing how the number of toothpicks increases

One group chose to focus on how the diagonal of the figure changed from one staircase to the next (see fig. 13.3b). In this instance one new "step" is added which increases the perimeter by two along the diagonal, one horizontal toothpick and one vertical toothpick. In addition, one new vertical toothpick is needed to increase the height of the staircase and one new horizontal toothpick is needed to increase the width of the staircase. In the diagram, dark squares are replaced by squares with circles in them by moving one square left. Previously uncounted sides are marked with slash marks.

Another group visualized the previous staircase inside the new one (see fig. 13.3c) and looked at which toothpicks that had been on the perimeter of the previous staircase were now inside the new staircase. They too saw that four new toothpicks were added to the perimeter. The arrows in figure 13.3c indicate how toothpicks on the previous staircase's perimeter are counted in the new perimeter.

Two other groups failed to reason inductively. Students in these groups analyzed numerical relationships between the staircase number and the perimeter. They found the number of toothpicks needed to make the perimeter of a staircase was four times as large as the staircase number ($P = 4n$). However, neither group was able to explain convincingly why the number of toothpicks in the perimeter of the staircases should increase by four from one staircase to the next. In essence, these groups did not answer the questions that were asked.

As groups present their work, the instructor highlights similarities and differences among the geometric arguments and discusses why such arguments are necessary to show that the pattern of adding four toothpicks will continue as the size of the staircase increases. The instructor then extends the students' work by introducing the concept of mathematical induction and asks students what aspects of their arguments use mathematical induction. The instructor's goals are to motivate students to see the need for proof by mathematical induction and to help students understand how their informal arguments contain all the necessary steps for proof by mathematical induction.

To help students make these connections, the instructor first presents a formal proof by mathematical induction that the sum of the squares of the first n natural numbers is given by the formula

$$\frac{n(n+1)(2n+1)}{6}.$$

Although this formula is not directly related to the Staircase Problem, the method used in the proof parallels the informal reasoning students use as they solve the Staircase Problem. In presenting this proof, the instructor makes extensive reference to the students' work with the Staircase Problem so that students see where each piece of the formal proof arises in their informal arguments. For example, when the basis step of the formal proof is discussed (verifying the formula for $n = 1$) the instructor asks students to determine their basis step in the Staircase Problem (showing that their rule holds for the first staircase). Similarly, during the inductive step of the formal proof the instructor highlights how comparable reasoning arises in students' informal arguments, namely, assuming the formula holds for the nth staircase and using that assumption to show that the formula holds for the $(n + 1)$th staircase.

This activity provides opportunities for students to grow both in their mathematical knowledge and in their appreciation of middle grades mathematics. To begin, students seem to have a greater trust in the validity of proof by mathematical induction after using it informally and intuitively to justify their patterns in the Staircase Problem. Students believe that they have shown why the staircase pattern continues on the basis of their arguments. Hence, the idea of assuming something is true for a particular value of n and using that assumption to show that the same statement holds for the next value of n does not seem as foreign to them.

In addition, because the Staircase Problem is based on a simple linear relationship, the students focus the majority of their time and effort on analyzing why the pattern continues to hold. Certainly other activities can be used to introduce the idea of proof by mathematical induction, but this task minimizes the amount of algebraic manipulation involved and is geometric in nature. Hence, students keep their sights set on their main goal of explaining their reasoning using the geometric figures and do not get bogged down trying to determine the pattern.

Finally, the Staircase Problem introduces students to an activity that can be used with middle grades children. College students benefit by experiencing activities that are accessible to students at a variety of levels. They start to see that different questions can be asked to elicit different mathematical responses. Specifically, the college students see how middle grades children's use of a recursive formula (for example, a Now-Next formula) is related to proof by mathematical induction.

CONJECTURING AND PROVING PROPERTIES IN A EUCLIDEAN GEOMETRY COURSE

Another course in which conjecture, reasoning, and proof play a primary role is Euclidean Geometry. One goal of this course is to help students move through the levels of the van Hiele model of development of geometric thought (Crowley 1987). The van Hiele model identifies five levels of development of geometric thought: visualization, analysis, informal deduction, formal deduction, and rigor. At the level of *visualization,* the student identifies, names, compares, and operates on geometric figures according to their appearance. For example, a student reasoning at this level might remember what a rectangle is because "it looks like a door," not because it has four straight edges and four right angles.

At the level of *analysis,* the student analyzes a figure in terms of its components and relationships among components. Properties of a particular shape are discovered empirically by methods such as folding, measuring, or using a grid. Relevant attributes are understood by the student and irrelevant attributes such as orientation are ignored.

At the level of *informal deduction,* the student logically interrelates previously discovered properties or rules by giving or following informal arguments. Sufficient conditions needed to "prove" that a figure is a particular geometric shape can be selected. A student reasoning at this level understands relationships among general classes of figures and can order classes of figures. For example, a student would recognize that a square is both a rectangle and a rhombus.

At the level of *formal deduction,* the student proves theorems deductively and establishes interrelationships among networks of theorems. The role of axioms, undefined terms, and theorems is fully understood. A student reasoning at this level can construct an original proof and recognizes and appreciates geometry as a formal mathematical system.

At the level of *rigor,* the student establishes theorems in different axiomatic systems and analyzes and compares these systems. Axioms themselves become the object of intense, rigorous scrutiny. A student reasoning at this level understands what happens to Euclidean geometry if the parallel postulate is not assumed to be true.

It has been the experience of the Euclidean Geometry team that many students begin the course somewhere between the levels of visualization and analysis. The course activities help students move to the level of formal deduction and, in some cases, to the level of rigor. The course covers Euclidean geometry of the plane, including transformational geometry, with forays into finite, spherical, and hyperbolic geometries in order to compare and contrast Euclidean concepts and axioms in several different geometries. Students study triangles and quadrilaterals by beginning very informally at the level of visualization (finding examples in their environment and print media) and analysis (for example, looking for properties of parallelograms). Activities are designed to surprise and challenge students and to introduce fundamental concepts concretely while moving students to a more abstract conceptual understanding. One such activity is discussed here.

In many high school geometry textbooks, questions asking students to characterize specific quadrilaterals by their diagonals are spread throughout the textbooks' treatments of quadrilaterals. In the Diagonal Exploration (see fig. 13.4), students begin with a list of possible relationships between diagonals of quadrilaterals and are challenged to determine which types of quadrilaterals arise from diagonals satisfying the specific criteria. They conclude the activity by defining the quadrilaterals by the relationship(s) between their diagonals and proving these definitions are equivalent to the more common definitions of the quadrilaterals.

The instructors notice that many students initially arrange the diagonals more specifically than the stated relationships. For example, students make the diagonals intersect at midpoints or arrange them to be perpendicular to each other when these relationships are not specified. They also are uncertain what to do when the diagonals do not intersect. Since students work in small groups, the instructor addresses specific misconceptions as they arise within each group. Students complete the activity by characterizing each of

Diagonal Exploration

1. Experiment with the diagonals of quadrilaterals using the materials provided. Position the diagonals, then connect their endpoints to create a quadrilateral. Try at least three different arrangements for each diagonal relationship listed. Determine all quadrilaterals that **MUST result** from each diagonal relationship. List those quadrilaterals in the appropriate cells.

Relationship between Diagonals	Noncongruent Diagonals	Congruent Diagonals
a. Diagonals are perpendicular bisectors of each other.		
b. Diagonals bisect each other.		
c. Diagonals are perpendicular (and they intersect).		
d. Diagonals are perpendicular and one bisects the other (and they intersect).		
e. Diagonals are perpendicular and one, if extended, bisects the other. (They don't intersect).		
f. Diagonals of *ABCD* intersect in point *E* such that $\frac{AE}{EC} = \frac{BE}{ED}$.		

2. Write a definition of each type of quadrilateral in terms of its diagonals:

   a. Isosceles trapezoid b. Kite c. Parallelogram d. Rhombus
   e. Rectangle f. Square g. Trapezoid h. Chevron (concave kite)

3. Which categories of quadrilaterals belong to other categories? Explain briefly. Create a tree diagram showing the hierarchy of quadrilaterals.

4. Choose one of the definitions you wrote in (2) and prove that it is equivalent to the standard definition of the quadrilateral.

Fig. 13.4. Diagonal Exploration for the Euclidean Geometry course

the quadrilaterals by their diagonals, arranging the quadrilaterals in a tree diagram showing their hierarchy, and writing a definition for each quadrilateral on the basis of its diagonals.

The Diagonal Exploration forces students to think more deeply about quadrilaterals than many have previously. Many are bothered by the unfamiliarity of the definitions by the diagonals and either spontaneously, or with the instructor's guidance, begin to question if the quadrilaterals defined by diagonals are indeed the familiar ones that they know. Though plastic straws are available to model diagonals and to guide students' intuition to conjecture what quadrilaterals must be formed, the question remains: how do you know that, for all possible size diagonals in the specified arrangement, you always get a certain type of quadrilateral? This question and uncertainty promotes the need to prove that the student-created definitions are equivalent to the more familiar quadrilateral definitions. As a result of this activity, students' understanding of the meaning of definitions, along with their proof-writing abilities, is sharpened.

The Diagonal Exploration also demonstrates the importance of axiomatic systems. When students begin proving that the alternative definitions are equivalent to the more standard ones, discussions about underlying assumptions arise. For example, where is this proof beginning? One of the proofs begins with an assumption that the diagonals of a quadrilateral are perpendicular and one diagonal bisects the other. This sets up the idea that in order to begin any proof, certain assumptions must be made, some of which might be hidden. For instance, given this statement about the arrangement of diagonals, almost all students assume the diagonals intersect without explicitly stating that fact. The assumption is implicit in their drawings. Students are led to question whether drawings accurately convey a mathematical proof. This leads to the idea that drawings often contain hidden assumptions and that some basic agreed-on assumptions (axioms) are needed on which to build theorems.

Students derive abundant benefits from hands-on explorations throughout the Euclidean Geometry course. Axiomatic systems and proof develop naturally as students analyze and make conjectures about triangles and quadrilaterals, devise precise definitions, and compare the properties of triangles and quadrilaterals in the Euclidean plane with those in the hyperbolic plane and on the sphere. Undefined terms, axioms, and definitions arise through discussions and explorations. Proof begins with informal deduction arising from the explorations, proceeds to formal deduction based on one of the geometric models, and progresses to rigor when students are able to write neutral geometry proofs without relying on a particular model. Finally, students once again see the complexity of geometric ideas that arise in school mathematics.

CONCLUSIONS

As the examples presented in this article illustrate, the inclusion of grades K–12 mathematics activities in core college-level mathematics courses enables students to acquire a deeper and more connected understanding of the mathematical content. In addition, these activities motivate prospective teachers as they realize that they need profound understanding of the mathematical concepts that they will teach in grades K–12 classrooms.

Finally, collaborating to include grades K–12 mathematics materials in the core courses has benefited faculty. Faculty involved in finding, adapting, and creating materials for these courses have come away with a better understanding of national standards and grades K–12 mathematics curricula, as well as a deeper appreciation for the level of mathematics present in exemplary grades K–12 materials. Participation in this project has also increased faculty communication within and among departments. This increased communication has strengthened a collegial and collaborative environment among faculty and has facilitated the compilation of a large collection of activities that can be used to teach mathematics and statistics courses.

REFERENCES

Conference Board of the Mathematical Sciences (CBMS). *The Mathematical Education of Teachers.* Washington, D.C.: Mathematical Association of America, 2001.

Crowley, Mary L. "The van Hiele Model of the Development of Geometric Thought." In *Learning and Teaching Geometry, K–12,* 1987 Yearbook of the National Council of Teachers of Mathematics (NCTM), edited by Mary M. Lindquist and Albert P. Shulte, pp. 1–16. Reston, Va.: NCTM, 1987.

Garfield, Joan, and Andrew Ahlgren. "Difficulties in Learning Basic Concepts in Probability and Statistics: Implications for Research." *Journal for Research in Mathematics Education* 19 (January 1988): 44–63.

Hiebert, James, and Thomas Carpenter. "Learning and Teaching with Understanding." In *Handbook of Research on Mathematics Teaching and Learning,* edited by Douglas Grouws, pp. 65–97. New York: MacMillan, 1992. (Available from National Council of Teachers of Mathematics, Reston, Va.)

National Commission on Mathematics and Science Teaching for the Twenty-first Century (NCMST). *Before It's Too Late: A Report to the Nation.* Jessup, Md.: Education Publications Center, 2000.

National Council of Teachers of Mathematics (NCTM). *Principles and Standards for School Mathematics.* Reston, Va.: NCTM, 2000.

Schoenfeld, Alan, Hugh Burkhardt, Phil Daro, Jim Ridgeway, Judah Schwartz, and Sandra Wilcox. *Balanced Assessment: Middle Grades Assessment.* White Plains, N.Y.: Dale Seymour Publications, 1999.

Usiskin, Zalman. "Teachers' Mathematics: A Collection of Content Deserving to Be a Field." *The University of Chicago School Mathematics Project Newsletter* 28 (winter 2000–01): 5–10.

14

Promoting Content Connections in Prospective Secondary School Teachers

Irene Bloom

THERE has long been agreement that prospective high school mathematics teachers need a substantial amount of study beyond the high school content they will teach. But it has not been so clear what that mathematics might be. Educators at Arizona State University have created a three-course sequence that aims to provide preservice secondary school teachers with a well-connected understanding of secondary school content and to guide them through the transition from thinking like students to thinking like teachers. In these three courses, they experience the structure of mathematics and how various content domains are related to one another and to the real world. Our goal is to prepare confident, flexible, knowledgeable future teachers who are at ease with concepts and are capable of asking probing questions that build inquiry communities where learners make sense of mathematics.

When Are We Going to Use This?

Our interviews with student teachers and beginning teachers revealed that most saw little or no connection between the upper-division courses they took and the mathematics they were now attempting to teach.

> *Beth:* What was missing for me were applications ... where we can use it in the classroom and what was the relevance of what we were learning with respect to the algebra one, algebra two, and so forth in high school.

The research reported in this manuscript was supported, in part, by grant no. 9876127 from the National Science Foundation and grant no. P336B990064 from the U.S. Department of Education. The opinions expressed are solely those of the author and do not reflect the opinions of the National Science Foundation or the U.S. Department of Education.

The Conference Board of the Mathematical Sciences (2001) acknowledged that few college students are able to make connections between the abstract mathematics they study in college and the high school mathematics they will teach. A recent graduate commented, "Like linear algebra—I mean it was really neat, I liked the math, but I can't really see using any of that stuff in my teaching." Interviews with other teachers from our program revealed that they wished they had taken some "lower level" courses so that they could refresh their skills. They reported that they had forgotten many of the formulas and algorithms they used in secondary school mathematics.

> *Andy:* It would be good to have even like a trig class or a college algebra class even just for a basis ... because that's where you're sitting at the high school level.

Those very skills—fresh or not—are inadequate when trying to construct rich, interactive classrooms where mathematical inquiry is to take place. Ball (1990) found that secondary school mathematics education majors were no better at explaining simple concepts, such as division, than were prospective elementary school teachers. My own students have found themselves speechless when asked questions such as, "Why is $x^0 = 1$?" or "Why does a negative number times a negative number equal a positive number?" They know how to use these facts, but they don't know how they are derived. This is not because they are incompetent mathematics students; it is simply because no one asked them before. As one student said, "It's just because. No one ever explained it to me; they just showed me how to do it. I never thought about it beyond that."

MATHEMATICS EDUCATION COURSES

Our university has focused on building students' understanding of the many processes and facts of secondary school mathematics, as well as learning to use best practices. Consequently, we have reformed our entire three-course sequence: a content course and two methods courses.

One methods course attends to the concept of function and the various ways it surfaces across the secondary school curriculum. The other methods course considers the topics of geometry, number, and data analysis. The approach of examining concepts in the context of good pedagogy has been fruitful in enabling preservice teachers to grapple with their own understandings while they concurrently explore methods of teaching.

The third course, Mathematics Content in the Secondary School, is the focus of the remainder of this article. It provides a platform to concentrate on secondary school mathematics from the perspective of future teaching and learning.

DOING MATHEMATICS

One of the ways we address content in the secondary school is to have students solve problems from a wide variety of secondary school mathematics resources including textbooks, teaching journals, contest materials, and others. Like many good mathematics students, our students come to us with much of their knowledge compartmentalized. They are unable to see or appreciate how high school topics are related. We design homework to include problems selected from different content areas. As a result, students are required first to determine the mathematical facts and concepts they need to proceed. Here is a typical homework assignment.

1) Find the area of a rhombus that has one side of length 10 and diagonals that differ by 4.

2) Consider the family of quadratic functions given by $f(x) = (m^2 - m + 1)x^2 - (2m)x + 1$ where m is any real number. For what values of m will $f(x)$ always be positive?

3) Solve the following system of equations:

   $x^2 + y^2 = 208$

   $xy = 96$

4) Graph the following equation and compute the area it encloses.

$$|2x - 10| + |5y - 10| = 20$$

High-achieving math students who excel at completing the exercises at the end of a chapter or section are greatly challenged when confronted with problems that are stripped of the familiar cues.

Kate: I'm so used to using whatever the book had. With these problems, sometimes I don't know where to start. I'm not used to looking at problems this way.

Manny: It's been so hard because we have been trained to answer questions that come from a particular chapter. We are not taught how to cross these subjects so it becomes difficult when we are trying to solve problems that come from many different areas.

Structuring the homework this way accomplishes several goals. Students begin to ask themselves important questions. What mathematics content is required to solve the problem? Where does this problem fit in the secondary school mathematics curriculum? What is the most efficient way to solve the problem? How can I make sense of this problem? During class, different students demonstrate their various solutions for the problems. If a student explains a problem in a way that some classmates do not understand, that student must make his or her reasoning clear. This is difficult at the beginning of the semester, but they improve quickly.

> *Neal:* I saw the problem as a straightforward system of equations; I was surprised how differently some people saw it.

Students gain experience in trying to follow someone else's reasoning, even if it does not match their own line of reasoning for that problem. As they become more adept at tackling problems, their confidence regarding the content begins to grow. They begin to feel more "expert," relying on their own knowledge and intuitions.

> *Carla:* [One] good thing I can say about seeing people work on a problem is that I questioned the validity of the solutions they came up with. I wasn't exactly convinced with some of the solutions I heard.

Classroom activities focus on difficult problems that often take more than one class period to solve. Students work in groups and share their solutions with the class. The activity in figure 14.1 (borrowed from Schoenfeld 1985, p. 21) was especially taxing.

You are given a fixed triangle *T* with base *B* as below. Show that it is possible to construct, with a straightedge and compass, a straight line that is parallel to *B* and that divides *T* into two parts of equal area.

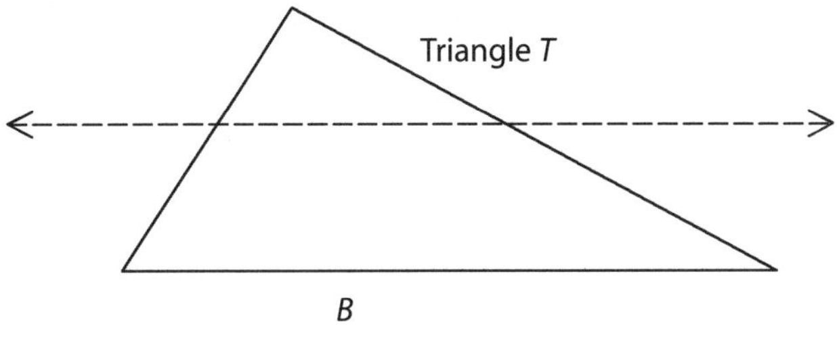

Source: Schoenfeld 1985, p. 21.

Fig. 14.1

Each group eventually solved the problem satisfactorily. Two distinct subgoals emerged as students attempted to solve this problem. One goal was to determine the height of the triangle created by the parallel line. This necessitated using properties of similar triangles to set up a series of proportions. See figure 14.2 for one group's solution.

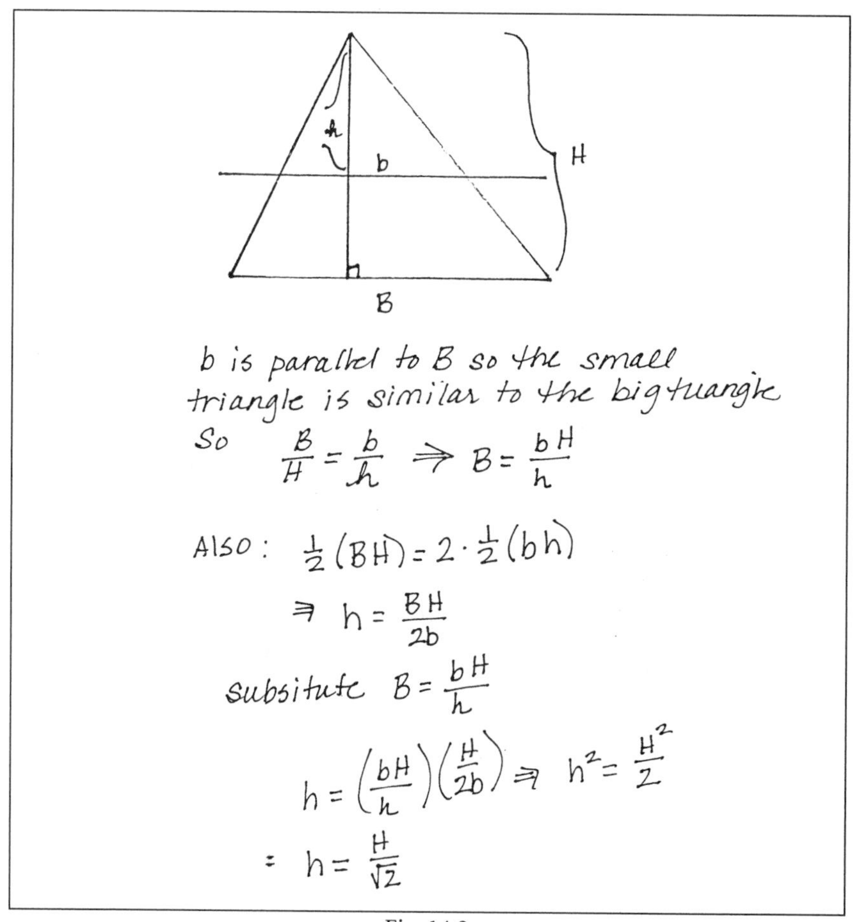

Fig. 14.2

The second goal involved unraveling the issue of how to construct the needed length. Each group assembled a unique path to the solution. One such construction is represented in figure 14.3.

Although it seemed to them that they "wasted" a good bit of class time on it, their subsequent reflections on the activity revealed that it was time well spent. As Yolanda stated, "I felt pushed from my comfort zone." They were forced to confront inadequacies in their knowledge base, frustration, and confusion. They also had to examine some of their beliefs about mathematics and problem solving that may have hindered their progress.

Tom: I think it is just implanted when you like go along, it's like either you can solve the problem in 5 or 10 minutes, other-

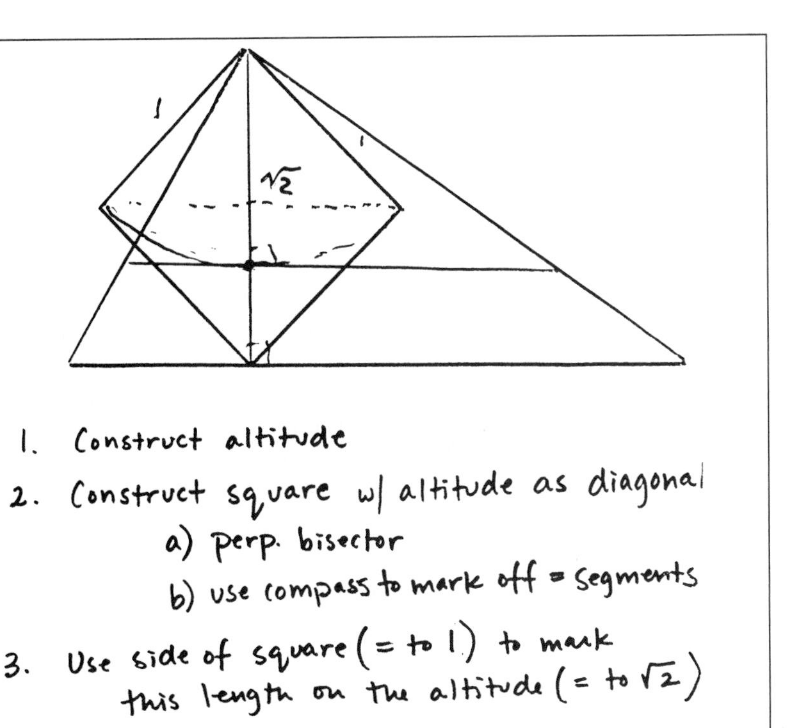

1
1
√2
1. Construct altitude
2. Construct square w/ altitude as diagonal
a) perp. bisector
b) use compass to mark off = segments
3. Use side of square (= to 1) to mark
this length on the altitude (= to √2)
4. Construct perp. thru mark made in #3

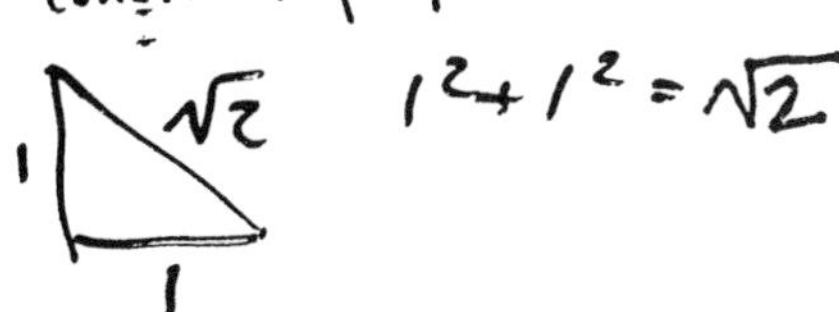

Fig. 14.3

> wise, it is too hard. And that's not the case, it is just that you really have to sit there and think about and use all that you know about the problem.

The students were asked why this particular problem was so difficult for them. The responses varied.

- It wasn't that the math was so hard. I just didn't think to apply any of it.
- Our group kept trying to figure it out with just geometry. When we realized we had to use proportions, it was trivial.
- I realize that I don't really understand proportions.
- I've never used a compass and a straight edge to solve a problem.
- I kept thinking that $h/\sqrt{2}$ couldn't possibly be right, so we kept getting stuck.

- I didn't really know how to start.

Students also participate in extended analyses of common word problems from the high school curriculum. Stanley and Callahan (2002) have developed a series of tasks for use with in-service teachers, and we have modified a subset of them for use with preservice teachers.

This is one of the Stanley and Callahan tasks we pose to students:

> Suppose that on a round trip you go at 30 mph on the way out and at 60 mph on the way back. What is your average speed?

Students were asked to solve the problem in groups and present their solutions to the class. In conjunction with these activities, we generally have a class discussion regarding difficulties that high school students may experience, common errors that they might make, and where the problem may surface in the high school curriculum. Figure 14.4 shows a typical solution.

We know that $D=RT$, therefore,
the rate is distance divided by time, $R=\frac{D}{T}$.

The time for the outbound trip plus the time for the return trip equals the total time using the average rate of speed, x.
The distance for the total trip is $2d$.

$$\frac{d}{30}+\frac{d}{60}=\frac{2d}{x}$$

$$\frac{2d}{60}+\frac{d}{60}=\frac{2d}{x}$$

$$\frac{3d}{60}=\frac{2d}{x}$$

$$3x=120$$

$$\underline{x=40}$$

Fig. 14.4

Prospective teachers are then asked to look at the problem in a more general way (see fig. 14.5). Numbers are replaced with parameters, and equations are manipulated in different ways to reveal some of the structure of the problem type, including that the average rate is the harmonic mean of the original rates.

Let d = distance, r_1 = rate of the outbound trip, and r_2 = rate of the return trip. Again, the time outbound plus the time to return equals the total time using the average rate.

$$\frac{d}{r_1}+\frac{d}{r_2}=\frac{2d}{x}$$

$$\frac{1}{r_1}+\frac{1}{r_2}=\frac{2}{x}$$

$$r_2x+r_1x=x(r_2+r_1)=2r_1r_2$$

$$x=\frac{2r_1r_2}{r_1+r_2}$$

Which is equivalent to $x=\dfrac{2}{\left(\dfrac{1}{r_1}+\dfrac{1}{r_2}\right)}$.

Fig. 14.5

In this example, students may be encouraged to compare the harmonic mean of the two speeds and the arithmetic mean and discuss why the arithmetic mean does not produce the correct answer, or they may investigate what kinds of limiting factors exist for the return speed on the basis of the structure of the problem. Figure 14.6 shows one future teacher's analysis.

In the final step, the students create problems that are isomorphic to the problem they explored and demonstrate that their problem contains the same basic structure as the original problem. Writing problems that are structurally equivalent, but contextually different, is a valuable skill for any teacher. The following is a problem one student wrote.

Suppose you earn the same amount of interest on 2 different savings accounts. One account pays out 4% interest per year and the other pays out 3% interest per year. What is your average rate of return per year?

He then demonstrated that the problem was structurally equivalent to the round-trip problem.

As a final project, students select a problem from a high school textbook, perform an independent extended analysis, and share their discoveries with

If I hold the first rate constant and I let the second rate grow, my average will never go above twice the first rate.

$$\lim_{r_2 \to \infty} \frac{2}{\left(\frac{1}{r_1} + \frac{1}{r_2}\right)} = 2r_1$$

So in the original problem, no matter how fast I make the return trip, my average will not exceed 60 mph.

Fig. 14.6

the class. When students make connections on their own, it has a tremendous impact. One pair of students selected a geometry problem that required finding the area of a trapezoid using at least two different methods. In their subsequent analysis and generalization, they ended up proving the Pythagorean theorem. They were very excited about this unexpected connection. Figure 14.7 shows a student's analysis of an area problem.

Breaking Down Barriers

As the semester progresses, the barriers that keep content areas compartmentalized start to break down and students begin to see mathematics as a continuum of concepts that are interrelated, rather than a set of isolated skills and formulas. They discover that certain problems resurface year after year in the curriculum sequence and can be solved with increasingly sophisticated skills. For example, consider various strategies for solving a problem that asks for the maximum of a quadratic function.

- Create a table of values and "zero-in" on the maximum.
- Graph the parabola and trace to, or estimate, the maximum point.
- Find the vertex of a parabola using the standard form or a formula.
- Find the vertex of a parabola by completing the square.
- Take the derivative and search for critical points by determining where the slope is zero.

A class discussion regarding such a problem would center on when these different strategies might be introduced in the curriculum, and in what contexts one method would be more valid or useful than another.

By the end of the course, future teachers start to see school mathematics as a well-connected web of major concepts and ideas. Discussions reflect recognition of the careful scaffolding that takes place as concepts build on one another. They report greater ease and confidence with the content as well as renewed enthusiasm for exploring mathematical ideas and problems.

> *Tom:* I think [my abilities have] gone through more of an openness to variety, whereas my problem solving had been more

Original Problem

Mr. Schultz doubled the area of his rectangular garden by adding a strip of new soil of uniform width along each of the sides. If the dimensions of the original garden were 10′ × 15′, how wide a strip did he add?

Solution to Original Problem

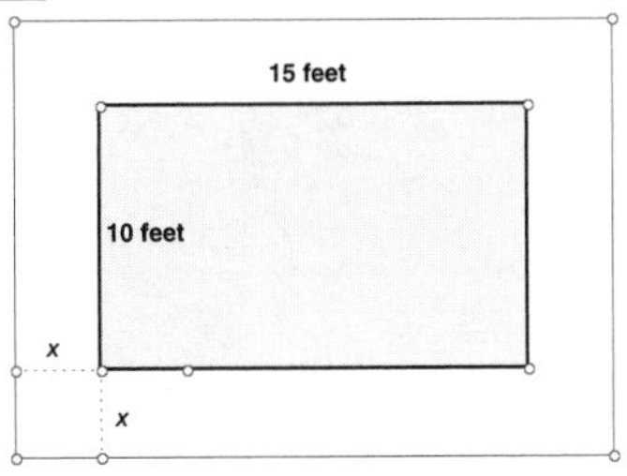

Let x equal the width of the new strip of soil.

- Double the area of the original. $2(150)$
- Set this equal to the new dimensions. $(15+2x)(10+2x)=300$
- Solve for x. $4x^2+50x-150=0$
- Use the quadratic formula. $x=\dfrac{-50\pm\sqrt{2500-4(4)(-150)}}{8}$
- Solve for x. $x=2.5$ or $x=-15$

We cannot have negative length, therefore $x=2.5$ ft.

Now Generalize the Problem:

Let a and b be the sides of the original figure, and x remain the width of the strip around the outside.

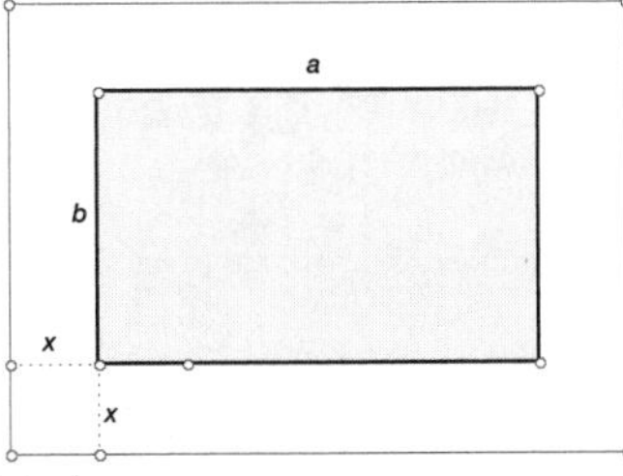

- Double the area of the original. $2(ab)$
- Set this equal to the new dimensions. $(a+2x)(b+2x)=2ab$
- Solve for x. $4x^2+x(2a+2b)-(ab)=0$

Fig. 14.7 *(continued on next page)*

narrow minded and I've not used my whole entourage of mathematics, ... by having to do a variety of problems [where] you don't know where they come from makes you use [all the] mathematics that you know.

Meanwhile, Heidi rediscovered why she wanted to be a mathematics teacher in the first place.

Heidi: It helped [me] remember why I loved math. We were about halfway through the semester when I realized—wow—I

- Use the quadratic formula.

$$x = \frac{-(2a+2a) \pm \sqrt{(2a+2b)^2 - 4(4)(-ab)}}{8}$$

At this point we can notice that the square root of the determinant, $\sqrt{(2a+2b)^2 - 4(4)(-ab)}$, will always be larger than $-(2a+2b)$. Therefore, to give a positive length, we will always *add* the square root of the determinant, as subtracting will yield a negative result.

- Simplify.

$$x = \frac{-(2a+2a) + \sqrt{(2a+2b)^2 + 16ab}}{8}$$

We notice the perimeter of the original figure is $P = 2a + 2b$, and that the area of the original figure is $A = ab$. By observing our derived formula, we see that we can write x in terms of the area and perimeter of the original figure.

$$x = \frac{-P + \sqrt{P^2 + 16A}}{8}$$

Therefore, given the dimensions of any rectangular figure, the width of the strip that needs to be added around the figure to double the area can be written in terms of the original area and perimeter.

Additional Explorations:

We tried to see if there was a maximum or minimum value for x:

- Take the derivative of $y = 4x^2 + x(2a+2b) - (ab)$. $y' = 8x + 2a + 2b$
- Set derivative equal to zero. $8x + 2a + 2b = 0$
- Solve for x and simplify. $x = \frac{-(a+b)}{4}$

Therefore there is a possible maximum or minimum at $x = \frac{-(a+b)}{4}$.

The second derivative is constant; therefore there are no inflection points. Thus, x continues to increase, meaning there is no maximum value, and since $(a+b) > 0$, there is no minimum value.

Finally, we looked at what would happen if he wanted to increase the area of the rectangle by a factor of n.

- Increase the area of the original by a factor of n. $n(ab)$
- Set this equal to the new dimensions. $(a+2x)(b+2x) = nab$
- Solve for x. $x_n = \frac{-P + \sqrt{P^2 + 16(n-1)A}}{8}$

Therefore we can find the width of the strip that needs to be added to any rectangle with dimensions $a \times b$ that will increase its area by a factor of n.

Fig. 14.7 (*continued*)

was really having fun in here. I want to take that with me to the schools [where] I will teach.

REFERENCES

Ball, Deborah Loewenberg. "Prospective Elementary and Secondary Teachers' Understanding of Division." *Journal for Research in Mathematics Education* 21 (March 1990): 132–44.

Conference Board of Mathematical Sciences. *Issues in Mathematics Education.* The Mathematical Education of Teachers, Vol. 11. Providence, R.I.: American Mathematical Society, 2001.

Schoenfeld, Alan H. "A Framework for the Analysis of Mathematical Behavior." In *Mathematical Problem Solving,* by Alan H. Schoenfeld, pp. 11–45. Orlando, Fla.: Academic Press, 1985.

Stanley, Dick, and Patrick Callahan. "The In-Depth Secondary Mathematics Institute." (Working papers, Texas Education Agency and the Texas Statewide Systemic Initiative of the Charles A. Dana Center at the University of Texas at Austin, 2002).

APPENDIX

Solutions to problems in the sample homework

1. Find the area of a rhombus that has one side of length 10 and diagonals that differ by 4.

Solution

Let $2x$ be the length of the shorter diagonal and let $2x + 4 = 2(x + 2)$ represent the length of the longer diagonal. Since the diagonals of a rhombus are perpendicular, we can then consider 1/4 of the rhombus as a right triangle with sides x, $x + 2$, and hypotenuse of 10. We can then use the Pythagorean theorem to determine x.

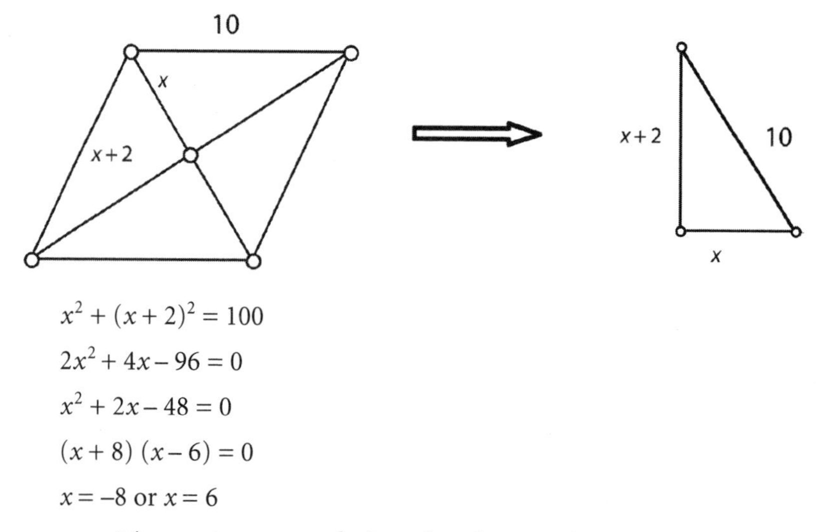

$$x^2 + (x + 2)^2 = 100$$

$$2x^2 + 4x - 96 = 0$$

$$x^2 + 2x - 48 = 0$$

$$(x + 8)\,(x - 6) = 0$$

$$x = -8 \text{ or } x = 6$$

$x = -8$ is an extraneous solution, therefore $x = 6$

Area of the 1/4 rhombus $= \frac{1}{2}(6)(8) = 24u^2$, so the area of the rhombus is $4(24) = 9u^2$.

---

2. Consider the family of quadratic functions given by $f(x) = (m^2 - m + 1)x^2 - (2m)x + 1$ where m is any real number.

   For what values of m will $f(x)$ always be positive?

Solution

In order for $f(x)$ to be always positive, the entire parabola generated by $f(x)$ must be above the x-axis, so we are looking for a parabola with a positive leading coefficient and with no real solutions (i.e., a negative discriminant).

Therefore $m^2 - m + 1 > 0$ and $(-2m)^2 - 4(m^2 - m + 1)(1) < 0$

Notice that $m^2 - \mathrm{m} + 1 > 0$ for all m, so we need only concern ourselves with the discriminant.

$$(-2m)^2 - 4(m^2 - m + 1)(1) < 0$$
$$4m - 4 < 0$$
$$m < 1$$

Therefore $f(x)$ will be positive for all $m < 1$.

---

3. Solve the following system of equations:

$$x^2 + y^2 = 208$$

$$xy = 96$$

Solution

Since $x^2 + y^2 = 208$ is a circle centered on the origin, and $xy = 96$ is a hyperbola centered on the origin, we see that they intersect in 4 places.

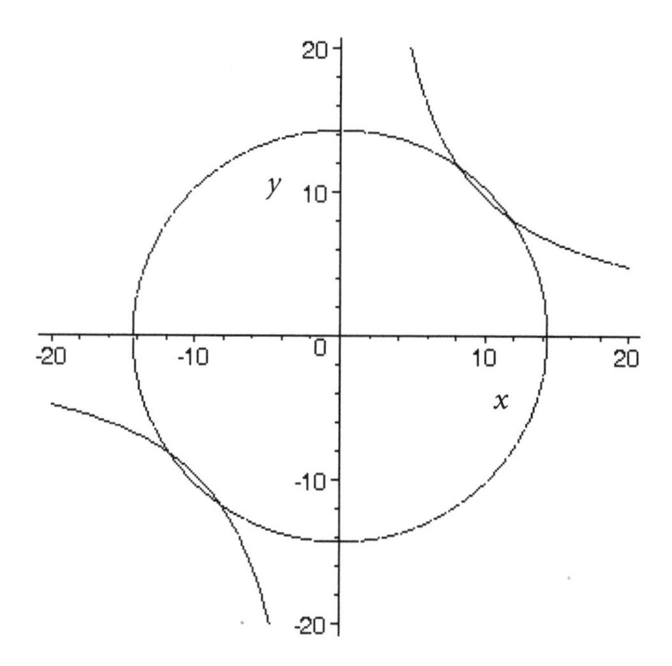

Solve both equations for y: $y = \pm\sqrt{208 - x^2}$ and $y = \dfrac{96}{x}$

Set the equations equal to one another: $\sqrt{208 - x^2} = \dfrac{96}{x}$

$x^2(208 - x^2) = 208x^2 - x^4 = 9216$

$-x^4 + 208x^2 - 9216 = 0$ factors into $(x + 12)(x - 12)(x + 8)(x - 8) = 0$

Therefore, $x = \pm 12,\ \pm 8$.

Solving for y yields four ordered pairs: (-12,-8), (12,8), (-8,-12), (8,12)

---

4. Graph the following equation and compute the area it encloses.

$$|2x - 10| + |5y - 10| = 20$$

Solution

The area enclosed by these 4 lines is a rhombus.

$$(2x - 10) + (5y - 10) = 20 \Leftrightarrow y = -0.4x + 8$$

$$-(2x - 10) - (5y - 10) = 20 \Leftrightarrow y = 0.4x$$

$$-(2x - 10) + (5y - 10) = 20 \Leftrightarrow y = 0.4x + 4$$

$$(2x - 10) - (5y - 10) = 20 \Leftrightarrow y = -0.4x + 4$$

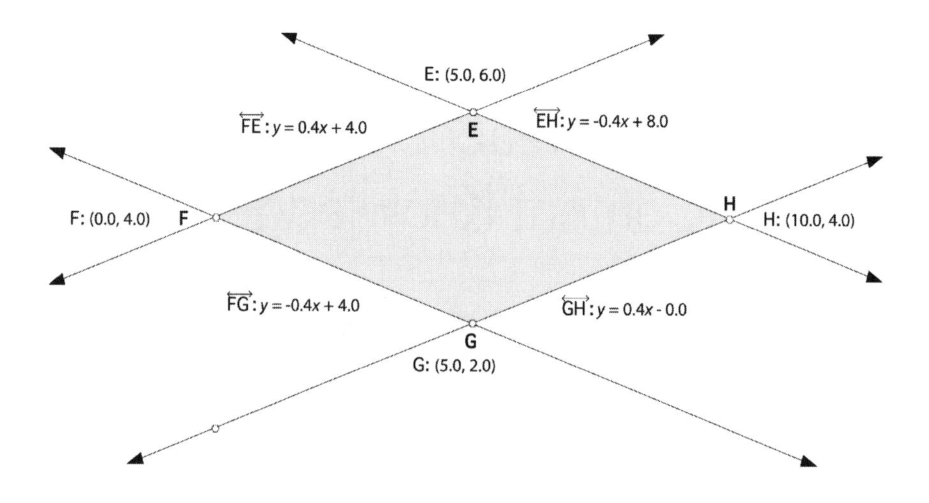

The vertices are: E = (5,6), F = (0,4), G = (5,2), and H = (10,4),

$$FH = \sqrt{(0-10)^2 + (4-4)^2}$$

or simply 10 – 0 (since the points lie on a horizontal line) = 10 units. $EG = 6 - 2$ (since the points lie on a vertical line) = 4.

As shown in problem #1, the area of a 1/4 of the rhombus is

$$\frac{1}{2}\left(\frac{1}{2}d_1\right)\left(\frac{1}{2}d_2\right) = \frac{1}{8}d_1d_2$$

making the area of the entire rhombus

$$4\left(\frac{1}{8}d_1d_2\right) = \frac{1}{2}d_1d_2 \; .$$

Therefore the area is $\frac{1}{2}(4)(10) = 20u^2$.

15

Facilitating Teachers' Growth in Content Knowledge

Kathleen Cramer

JUST as learning evolves over time, so does our learning to teach. Courses grow and change as instructors learn more about students' thinking and how students respond to different tasks, questions, assessments, and other aspects of teaching. In this article, I share what I have learned about teaching mathematics courses for elementary school teachers over several years and under three different National Science Foundation projects. I have learned a great deal about elementary school teachers' mathematical thinking, and this knowledge continues to grow and influence the content learning experiences I provide elementary school teachers.

I have developed and taught three content courses for teachers: Geometry, Number and Numeration, and Functions and Proportionality. In this article, I provide an overview of the Functions and Proportionality course and one group of teachers' reactions to that particular course, in order to offer readers a picture of the experience teachers in these three projects have had learning mathematics. Although I have taught the functions course to ten different groups of elementary school teachers, I will draw on examples from one of the Functions and Proportionality classes to highlight points made. This group included twenty-three certified elementary school teachers representing grades K–5 classrooms and special education. They took the class as a staff development opportunity independent of other teacher enhancement activities. Their teaching experience ranged from one year to thirty years, with an average of seven to eight years. Eight of the twenty-three teachers had been teaching less than four years.

This research was supported in part by the National Science Foundation under grant no. NSF-ESI-9454371. Any opinions, findings, conclusions, or recommendations expressed herein are those of the author and do not necessarily reflect the views of the National Science Foundation. The author is indebted to Seth Leavitt, a middle school teacher who has assisted in teaching courses described in this article. His insights and feedback have been invaluable.

AN OVERVIEW OF CONTENT COURSES

The pedagogical model used to guide the development of the Functions and Proportionality course and the other two content courses reflected the National Council of Teachers of Mathematics *Standards* vision of what a mathematics classroom could become. Whether learners are children or teachers, many characteristics are shared.

- Mathematics content is embedded in problem settings; learners collect data, generate hypotheses, and verify conjectures.
- Learners work in small groups to optimize the opportunity for discourse.
- Questions are posed to help learners construct mathematical knowledge.
- Learners' language (oral and written) plays an important role in facilitating the transition from problem solving and exploration to formal mathematical abstractions.
- Connections within and among mathematical topics are emphasized.
- Technology use is integrated into the daily activities of the course.

Rather than being survey courses, these courses were designed to address in depth a smaller number of topics. A conscious effort was made to make connections among topics within a course and topics across courses.

The Functions and Proportionality course was organized around the exploration of linear, quadratic, and exponential functions. The first part of the course examined linear, quadratic, and exponential functions in a variety of ways: concrete models, tables, graphs, algebraic rules, and context. Connections among different representations were emphasized throughout the lessons. Linear functions were then examined in detail, again with the focus on multiple representations. The last part of the course looked at proportionality. Teachers used their understanding of functions, and in particular linear functions, to understand what was special about proportional situations. As the mathematical characteristics of proportional situations were developed, teachers used previous learning to develop an understanding of this topic.

This content was organized into ten lessons of 3.5 hours each. The content taught was embedded in problems and leading questions that guided teachers to insights, patterns, and generalizations. The problems within each course built on one another to form a coherent whole. Lesson plans have been developed to describe the flow of each day's activities. Writing the lessons has been an ongoing process, so teachers' insights, difficulties, and solution strategies can be noted in the plans. Figure 15.1 shows a sample lesson plan from the Functions and Proportionality course that is representative of the lesson plans for each of the courses. (More information on the course, copies of the actual problem sets for this lesson, and additional sample lessons can be found on the Web at education.umn.edu/rationalnumberproject.) Similar activities have been adapted for use with middle school learners (Cramer 2001).

The Lesh translation model guided the development of all three courses (Lesh 1979). This model suggests that a deep understanding of mathematical ideas is obtained by experiencing mathematics through multiple representations (context, manipulatives, pictures, verbal symbols, written symbols) and by translating among the different representations. A concerted effort was made to find concrete representations for the mathematics studied as well as realistic contexts and pictorial models. For example, teachers explored functions by solving problems involving geoboards, pattern blocks,

Functions & Proportionality

Lesson 1 / Overview

Students will explore concrete examples of linear and quadratic functions. They will build tables, and explain, using informal language, patterns observed. Function rules will be recorded algebraically. For linear functions, two rules will be recorded.

Materials

POD: Transparency of POD 1

Student Pages:

Number Patterns from Cutting String; Patterns with Squares; Number Patterns with Equilateral Triangles; Practice Building Tables

Other Materials:

string, scissors, pattern blocks; transparency of course goals; transparency of Summary Tables

Teaching Actions

Problem of the Day (POD): What are the last two digits of this number written in standard form: 7^{9999}? Answer is 43.

Patterns observed:

$7^1 = 07$	$7^5 = 07$
$7^2 = 49$	$7^6 = 49$
$7^3 = 43$	$7^7 = 43$
$7^4 = 01$	$7^8 = 01$

One possible strategy: $9999 \div 4 = 2499$ remainder 3. All exponents with the same remainder (after dividing by 4) have the same last two digits. Expect a variety of strategies.

Large Group Introduction

1. Start class with a discussion of the role patterns play in mathematics. Much of what mathematicians do is to look for regularities in real and mathematical worlds and often use symbols to describe generalizations that can be made.

Example: 1000 years before Pythagoras, the Babylonians observed the relationship among sides of a right triangle.

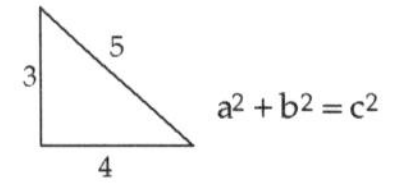

2. Explain that today and during the rest of the course they will explore patterns and regularities from different explorations, and they will find different ways to record their observations.

Comments

1. After completing the POD, share with students the class goals using transparency of goal sheet found in teacher's guide. Students then complete their self reflection on 3 of the course goals noted before doing the first large group activity. This form also is found in the teacher's guide.

2. Each class starts with a **Problem of a Day**. This is to engage students as they enter class.

Each class ends with students completing in their groups the **Big Idea** form. They are to write what they consider to be the important ideas developed in that day's lesson.

After Day 1 the **Big Idea** forms are used to review mathematics developed from the previous day. After discussing the POD, the instructor reads the forms and adds to them as needed to highlight the important ideas and connections made the previous day.

Each class then proceeds with the new topic for the day. This is done with large group introduction of the exploration that is completed during small group work. Large group discussion ends that sequence. This sequence may be repeated after the break.

Fig. 15.1. Sample lesson plan (*continued on next page*)

Teaching Actions

3. Set the stage for the string activity. Take a piece of string and cut it. Observe how many pieces there are. Ask: If I do that again, but cut twice, how many pieces of string will I have? Cut and verify guess. Complete a table and describe several patterns in the table.

Ask for patterns that could be used to extend the table to 10, 15, 100 cuts. Ask for the relationship to find the # of pieces given the # of cuts; the # of cuts given the number of pieces. **[p = c + 1; c = p - 1]**

# of cuts	0 1 2 3 4 5 6
# of pieces	1 2 3 4 5

Small Group Work

5. Have students complete **Number Patterns from Cutting String**. As students work, observe the types of patterns that they describe. If they are only recording the function rule, ask them to describe at least three different patterns observed in the table.

Large Group Discussion

6. Share patterns observed from conditions 1, 2 and 3. Don't limit the discussion to just the function rules. Record function rules algebraically connecting the use of variables with their informal description of the rules.

For example: Condition 1 - "To find the number of piece: you double the number of cuts and add one" **[p = 2c + 1]**

Consider the second rule for each table.

For example, for condition 1, write the rule predicting th number of cuts given the number of pieces:

If the **number of pieces equals two times the number c cuts plus one** then the **number of cuts equals the numb of pieces minus one divided by two** **[c = (p - 1) ÷ 2]**.

Ask how the two rules are related.

Comments

3. Students should observe as many patterns as possible using their informal language prior to recording patterns algebraically.

While the function rules are the patterns you will highlight, it is important to show how other patterns can be used to extend the table.

# of cuts	0 1 2 3 4 5 6
# of pieces	1 2 3 4 5

Students may note that if the # of cuts is even, the # of pieces is always odd; that the number o pieces is increasing by one as the # of cuts increases by one.

4. Discuss which patterns can be used to extend the table. Talk about how the function rule is the most efficient one to use when extending the data to 100 cuts.

5. Folding instructions for the three string problems:

Condition 1: string is folded in half before cutting.
Condition 2: string is looped over the scissors and then cut. The number of loops increases each time, not the number of cuts.
Condition 3: string is folded in half and tied at the end; the double string is looped over the scissors and then cut. The number of loops increases each time, not the number of cuts.

6. Students often need clarification on how to "loop" the string around the scissors for page 2 of the string activity.

7. The function rules are:

Cond # 1: p = 2c+1; c= (p-1)/2
Cond # 2: p = L + 2; L = p - 2
Cond # 3: p = 2L + 2; L =(p-2)/2
Extensions: p = 2L + 3; p = 3L + 4; p = 4L + 5

Teaching Actions

Break

Small Group Work

7. Continue the exploration of patterns by completing in groups the following two activity sheets; **Patterns with Squares** and **Number Patterns from Trains of Equilateral Triangles.**

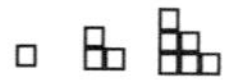

8. The function rule for **Patterns with Squares** is as follows:

Area = sum of the numbers 1 to t (t= the train number)

Note: This rule will be difficult to record algebraically; verbal explanation of what to do to find the answer is enough.

Perimeter = 4t (t = train number) ; Train # = P/4

10. **Number Patterns from Trains for Equilateral Triangles**

Students may need to be reminded that the unit of area for both parts of this activity is the green triangle.

Area = triangle number squared (A = t^2; (t = sq root of A)

Extension problem: Area = two times the rhombus number squared. **(A = $2r^2$; r = rhombus number).**

Large Group Discussion

11. Look for constant differences in all examples examined today. Use a transparency of the Summary Tables page. Sort problems into two groups by looking for constant differences in the tables. Match examples in each group with their functions. Note the differences in rules. Ask: Which group includes rules with only simple arithmetic? Which group includes rules with an exponen of two?

12. **Practice Building Tables** is provided if needed. Her students build tables given the rules and guess the rules, given the tables.

13. End class by completing the Big Idea form.

Comments

6. Students encounter examples of quadratic rules in the next two activities.

Patterns with Squares:
Area formula reflects the rule: y = n(n+1)÷2. This is Gauss' rule for adding consecutive numbers. You wil need to provide this rule.

You may want to model this rule using tiles:

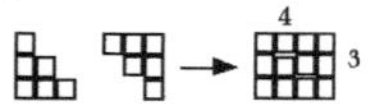

sum of numbers 1 to 3 = 3(3+1)/2

For this example it is important to look at patterns across the table and compare those patterns with ones in other tables.

train #	1	2	3	4	5
area	1	3	6	10	15

2 3 4 5
1 1 1

7. For quadratic rules, it takes two differences to reach a constant. For linear rules, a constant difference occurs the first time.

train #	1	2	3	4	5
perim	4	8	12	16	20

4 4 4 4

Fig. 15.1. Sample lesson plan *(continued)*

and Cuisenaire rods. They solved problems embedded in science contexts and real-life situations like converting between Celsius and Fahrenheit or finding the function relating wholesale and retail prices. In each activity, teachers were asked to describe their solution processes and reasoning. Teachers' informal language was the connection among manipulative models, contexts, pictures, and symbolic representations for the mathematical ideas.

Each class began with a problem of the day (POD) related to the day's activity or the previous day's lesson. The POD from the first day of the functions course is described in the sample lesson plan. After fifteen minutes or so, teachers shared their strategies and solutions. The instructor then introduced the day's first problem. Teachers worked in groups of their own choosing; periodically during the class, the instructor led large-group discussions of the problems.

Each class ended with a Big Idea form. In groups, teachers reflected on what they considered to be the important ideas learned in that session. These forms were used at the beginning of the next class to review the previous class's work. The instructor read them and then added her own observations to the points made.

Teachers' Responses to the Functions and Proportionality Course

At the end of the course, teachers were asked to reflect on the four statements in table 15.1. In a typical group, mean scores ranged from 4.3 to 4.8, with the highest score consistently connected to the teacher's style.

Teachers also were asked to comment on what was the most valuable part of the course. The majority of their responses can be grouped into three categories: pedagogy/teacher style, group work, and learning content. The following are some examples.

Pedagogy/teacher style

- Connecting the real-life situations to the development of function rules. Also I enjoyed the clear, logical way you could figure out new functions. The instructor was very clear and reassuring that the new learning would fall into place.

Group work

- I enjoyed working with my coworkers and seeing how they solve problems.

TABLE 15.1
Teacher Reflection Items

Circle the most appropriate response.

1. My knowledge and understanding of this mathematical topic has increased.

   (1) strongly disagree (2) disagree (3) neutral (4) agree (5) strongly agree

2. I am more aware of the importance of this topic as a critical component of mathematics.

   (1) strongly disagree (2) disagree (3) neutral (4) agree (5) strongly agree

3. I appreciated the teaching style of the instructor of these sessions.

   (1) strongly disagree (2) disagree (3) neutral (4) agree (5) strongly agree

4. My interest in learning more challenging mathematics has increased.

   (1) strongly disagree (2) disagree (3) neutral (4) agree (5) strongly agree

Learning content

- Learning more in-depth about the material and beginning to see connections to the mathematics I teach in my classroom.

Teachers' reactions showed that they appreciated the learning environment that emphasized group work, problem solving, the use of manipulatives, and real-world examples. Teachers' comments also showed the importance of attending to their feelings about learning mathematics. An instructor for content courses for elementary school teachers must realize that creating a safe, judgment-free environment is essential if teachers are going to learn. As one teacher stated about the instructor, "You are really patient—never shaming. Thanks from someone who struggled."

REFLECTING ON TEACHING MATHEMATICS TO TEACHERS

Pedagogy

There is a high degree of "mathematics phobia" among teachers who have taken this course, even though they volunteered to participate. Because of the high levels of anxiety, the instructor needs to be sensitive to how mathematics is taught and assessed. The universal, positive reaction to the courses can be attributed in part to the pedagogy that models a *Standards*-based vision of teaching and learning mathematics. The traditional model of teaching by telling is not appropriate for this population. The characteristics of *Standards*-based teaching that helped teachers overcome their anxiety about being in a mathematics class follow.

- Early use of concrete models to develop initial understanding of functions made these ideas accessible to teachers. From the first day, teachers felt successful, and these feelings helped sustain them when the content became more abstract.
- Science and realistic, contextual problems followed concrete examples; they provided teachers the opportunity to apply and extend their understanding of functions.
- Group work around problems provided a reason for teachers to work together cooperatively, building a sense of community within the class.
- Courses were carefully constructed, building connections among activities. Teachers could see their understanding of mathematics deepen as they moved from one activity to another.
- The instructor's role during the class varied. The instructor set the stage for the explorations. During group work, the instructor observed the groups, raising questions to help groups overcome misunderstandings. She helped teachers make connections with their previous work and built on teachers' ideas, extending the mathematics when appropriate.
- The instructor needed to be the teachers' advocate. At times when the content became particularly challenging, the instructor needed to review with the teachers what they knew and understood, reminding them how much they had learned.
- The instructor needed to balance cognitive demands of the content with the affective demands of working with this particular group of learners. Sometimes more time was needed on a particular topic than planned, and different problems or practice needed to be constructed to meet different needs of students. Despite well-planned and articulated lessons, adaptations needed to be made.
- The instructor was judgment-free, working with the learners from where they were both in their understanding of the content and how they felt about being learners of mathematics.

Assessing Content Knowledge

Assessment decisions should take into account teachers' feelings. Teachers realize that they do not know the mathematics they should know. They come to class open and at times embarrassed about their lack of knowledge. Although measuring content knowledge at the beginning of the class is important, if the instructor wants to measure growth and if teachers want to monitor their own growth, the teachers' feelings on that first day of class should be taken into consideration when choosing how to collect the data. From experience, I have learned that written pretests do not provide data valuable enough to warrant the level of anxiety it produces among the teachers. Teachers were anxious as they started the mathematics class; asking them to respond to a "test" on the first day of class only added to their anxiety, which in some instances lasted the entire course. I now ask teachers to communicate in writing their current understanding on a subset of the course goals. For the functions class, teachers responded to three of the six course goals. First, they were asked to describe connections among tables, graphs, and algebraic expressions for linear, quadratic, and exponential functions, noting differences among these three types of functions. Second, they were asked to describe connections between slope and y-intercept and to use these connections to graph linear functions or discover a function rule. Third, they were asked to list mathematical characteristics of proportional situations and to describe how to use them to distinguish proportional from nonproportional situations.

The instructions for this preassessment are worded to make it as nonthreatening as possible: "At the end of this course you will be taking a written assessment to document your mathematics knowledge of the content taught in class. You also will have the opportunity to develop a reflective journal communicating your knowledge of the course content. As you embark on this course, please reflect on the three highlighted goals by describing what you currently know about these mathematical areas. If you feel you currently have little understanding of the topic, just leave it blank."

Lack of Content Knowledge

Experience with these preassessment writings has shown that teachers' understanding of the course content is limited when they start the course. Although one would expect elementary school teachers to have some understanding of these algebraic ideas from their previous studies, in general teachers have reported very little content knowledge about these three types of functions and their representations or about proportionality.

Teachers' Content Knowledge at the End of the Course

Teachers in these courses learned a great deal about the topics taught. By the end of the course most teachers were able to distinguish among three types of functions by identifying patterns in tables, by examining characteristics of function rules, and by looking at the characteristics of their graphs. They could construct realistic examples for each type of function as well as solve realistic problems involving these three types of functions. They knew a great deal about linear functions in particular, and they were able to make connections among slope and *y*-intercept and their different representations: concrete, pictorial, graphical, and tabular. They were able to see proportionality as a special case of a linear function and solve problems involving proportionality using multiple strategies. After working with more than two hundred teachers, I know that these mathematical ideas are accessible to them and high expectations can be set.

To document their knowledge at the end of the course, teachers participated in two activities, a reflective journal and a written posttest. For their journal, they communicated in writing their understanding at the end of the course of the three goals they wrote about on the first day. Students completed this assignment at home and handed it in on the last class day. This assessment activity was presented as follows: "Document your growth of mathematical knowledge by reflecting on the three course goals [that is, say to yourself, "What do I know?"]. Write a summary for these three objectives that shows your understanding of each one. You may want to include and/or reference samples of class work. The purpose of this assignment is for you to communicate to me in your own words your mathematical knowledge." Teachers have consistently put forth a great deal of effort and time in completing these journals.

Responses to this assignment have been positive. For example, one teacher wrote the following at the start of her journal: "The reflections that follow were written right after each of the topics was covered. They were a useful way for me to crystallize my thinking about what I had just experienced and learned. They were helpful for me, also, as I reviewed for our final assessment."

The amount teachers wrote for each journal entry varied. Some were quite concise and conveyed a depth of knowledge on one page, whereas others wrote several pages for each objective. Figures 15.2 and 15.3 show samples that are representative of teachers' writings; these refer to the first objective that dealt with sorting through differences among three types of functions.

My task in reading the journals was to examine them for evidence of understanding and misunderstanding. Teachers received written feedback; I commented on positive aspects of their work, but I also pointed out any errors or lack of depth in their journals. Again, I knew I needed to be careful of teachers' feelings, acknowledging to myself that there would be differences among the teachers in content growth and levels of understanding. My goal was to make the writing experience productive for all teachers and to provide them with feedback that showed their growth but also informed them when their understanding lacked important ideas or connections or showed specific misunderstandings.

Common misunderstandings became evident in reading this group's journals. From their writings on the first objective, I realized that a few teachers

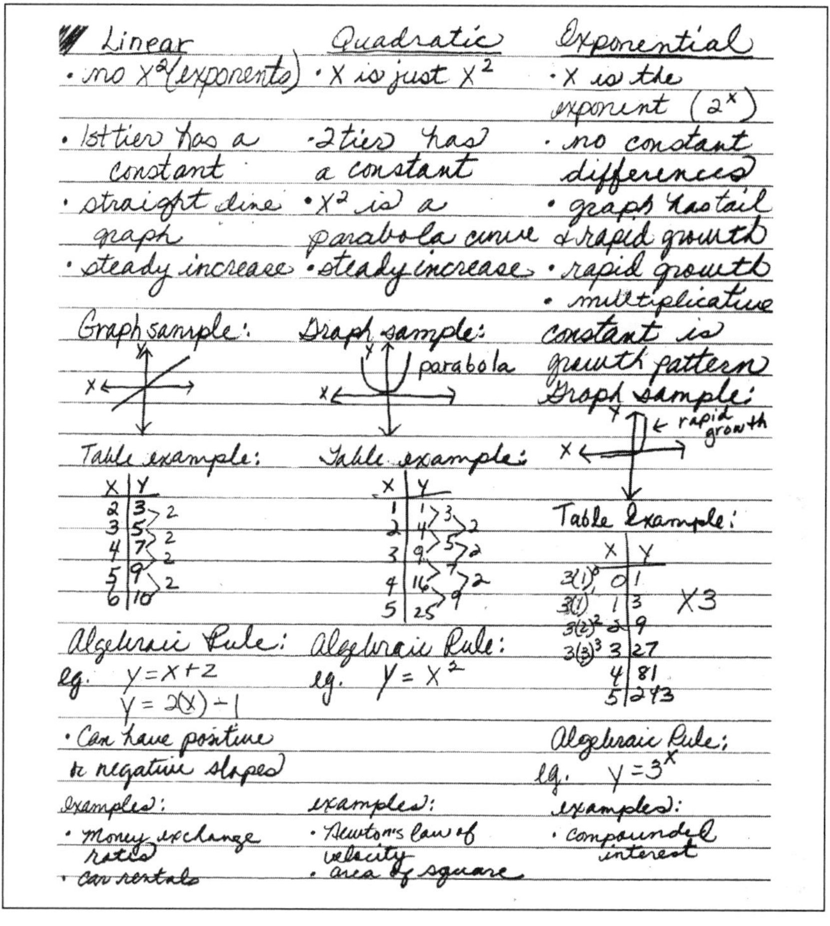

Fig. 15.2. Sample 1: teacher's journal entry at the end of the course

still confused terminology. For example, a teacher confused independent and dependent variables; in other instances, teachers mislabeled a quadratic function as exponential.

Overgeneralization of certain ideas was evident in a few journals. For example, a few teachers' writings implied that there were only three types of functions as opposed to having studied three examples of functions (see fig. 15.3). Some teachers overgeneralized about the rapid growth inherent in quadratic and exponential functions. Students would have benefited from examining and comparing a greater variety of examples to avoid misunderstandings about the growth of exponential and quadratic functions and their graphs. Course revisions reflect efforts to avoid these mistakes among teachers in future classes. For example, a new problem set is now included just

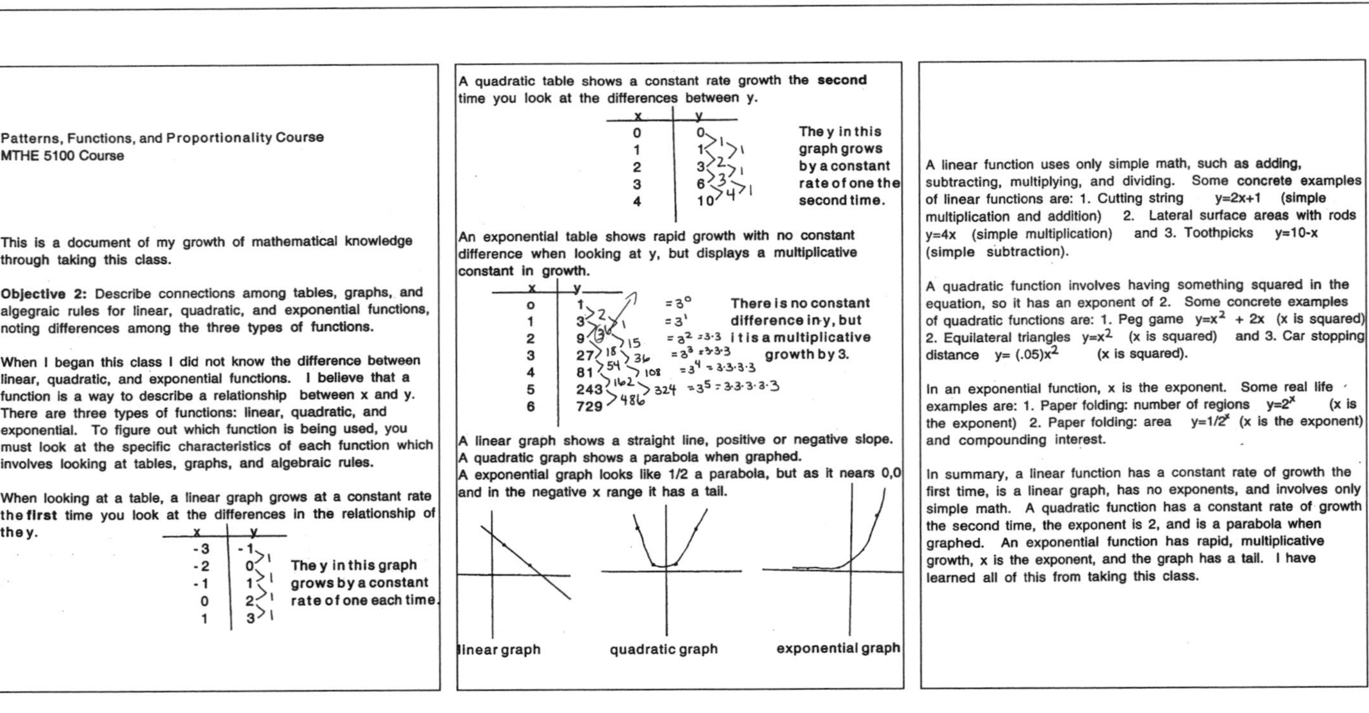

Patterns, Functions, and Proportionality Course
MTHE 5100 Course

This is a document of my growth of mathematical knowledge through taking this class.

Objective 2: Describe connections among tables, graphs, and algegraic rules for linear, quadratic, and exponential functions, noting differences among the three types of functions.

When I began this class I did not know the difference between linear, quadratic, and exponential functions. I believe that a function is a way to describe a relationship between x and y. There are three types of functions: linear, quadratic, and exponential. To figure out which function is being used, you must look at the specific characteristics of each function which involves looking at tables, graphs, and algebraic rules.

When looking at a table, a linear graph grows at a constant rate the **first** time you look at the differences in the relationship of the y.

x	y
-3	-1
-2	0
-1	1
0	2
1	3

The y in this graph grows by a constant rate of one each time.

A quadratic table shows a constant rate growth the **second** time you look at the differences between y.

x	y
0	0
1	1
2	3
3	6
4	10

The y in this graph grows by a constant rate of one the second time.

An exponential table shows rapid growth with no constant difference when looking at y, but displays a multiplicative constant in growth.

x	y	
0	1	$=3^0$
1	3	$=3^1$
2	9	$=3^2=3\cdot3$
3	27	$=3^3=3\cdot3\cdot3$
4	81	$=3^4=3\cdot3\cdot3\cdot3$
5	243	$=3^5=3\cdot3\cdot3\cdot3\cdot3$
6	729	

There is no constant difference in y, but it is a multiplicative growth by 3.

A linear graph shows a straight line, positive or negative slope.
A quadratic graph shows a parabola when graphed.
A exponential graph looks like 1/2 a parabola, but as it nears 0,0 and in the negative x range it has a tail.

linear graph — quadratic graph — exponential graph

A linear function uses only simple math, such as adding, subtracting, multiplying, and dividing. Some concrete examples of linear functions are: 1. Cutting string y=2x+1 (simple multiplication and addition) 2. Lateral surface areas with rods y=4x (simple multiplication) and 3. Toothpicks y=10-x (simple subtraction).

A quadratic function involves having something squared in the equation, so it has an exponent of 2. Some concrete examples of quadratic functions are: 1. Peg game $y=x^2+2x$ (x is squared) 2. Equilateral triangles $y=x^2$ (x is squared) and 3. Car stopping distance $y=(.05)x^2$ (x is squared).

In an exponential function, x is the exponent. Some real life examples are: 1. Paper folding: number of regions $y=2^x$ (x is the exponent) 2. Paper folding: area $y=1/2^x$ (x is the exponent) and compounding interest.

In summary, a linear function has a constant rate of growth the first time, is a linear graph, has no exponents, and involves only simple math. A quadratic function has a constant rate of growth the second time, the exponent is 2, and is a parabola when graphed. An exponential function has rapid, multiplicative growth, x is the exponent, and the graph has a tail. I have learned all of this from taking this class.

Fig. 15.3. Sample 2: teacher's journal entry at the end of the course

before exponential functions are introduced. This new problem set asks teachers to explore the effect the coefficient in a linear or quadratic function has on the growth of each function and to compare both functions as the coefficients are varied. Each time these courses are taught, new insights into teachers' thinking are revealed. The journals have proved to be an excellent tool for uncovering teachers' misunderstandings.

Although much of the work in class was done cooperatively, teachers were asked on the last day of class to complete a written final examination individually. There was still some anxiety at this time, which was why the test was not timed, and teachers used their notes and final journal as they completed the six-item test. All students were expected to complete four of the six items successfully, since these examples were similar to those done in class. The other two questions offered the teachers the opportunity to transfer their knowledge to new situations. They allowed teachers to demonstrate a depth of understanding the other four problems would not do; they also allowed teachers to demonstrate their ability to communicate their understanding in writing. In both instances, teachers could solve the problems in more than one way.

One of the higher-order questions follows.

> A ladder has 100 steps. On the first step sits 1 pigeon; on the second, 2; on the third, 3; and so on up to the hundredth step. How many pigeons in all?

A possible solution followed the problem:

> I made a table. On step one there would be 1 pigeon. For two steps there would be $1 + 2 = 3$ pigeons. For three steps there would be $1 + 2 + 3 = 6$ pigeons. I did this up to 4 steps. To find the total number of pigeons for 100 steps I said to myself: if 4 steps equal 10 pigeons, then 100 steps would be 25 times that. It would be 250 pigeons.

Teachers needed to determine if that strategy was reasonable and to explain their thinking.

This task provided teachers with the opportunity to choose among several big ideas developed in the course to solve the problem. The scenario reported one person's thinking, which assumed that the growth in the total number of pigeons was linear, and more specifically, proportional. This was incorrect. Because the increase in the total number of pigeons was not the same between every pair of steps (growth in the total number of pigeons was not constant), the relationship between step and total number of pigeons was not linear, and therefore could not be proportional. The number relationship embedded in this task was, in fact, quadratic. A rubric was used to sort teachers' responses to this problem. Twenty-one of twenty-three teachers answered the problem correctly, with eighteen giving a correct answer with a reasonable explanation. These explanations reflected four different strategies: estimation, checking the given strategy with another example, considering the type of growth in the table, and making connections to proportionality. Table 15.2 highlights examples for each category.

Teachers used a variety of strategies to solve the pigeon task. This is not surprising, because throughout the course teachers were exposed to many different problem situations and were encouraged to develop their own solution strategies. During the course participants had worked in small groups and presented their strategies to the entire class. This atmosphere of exploration and sharing allowed participants the freedom to construct a solution strategy that was meaningful to them.

Though there were differences in how teachers solved the pigeon problem, the strategies can be related to ideas developed in class. During the course, teachers were asked to identify a variety of patterns and relationships in data tables, with growth of the dependent variable and the function rule as the two most important ideas. But teachers were consistently asked to look for other patterns as well. To solve this task, some teachers did look at growth of the dependent variable to conclude that the answer given was unreasonable. Others identified different patterns within the data table to conclude the strategy used was inappropriate. For example, teachers identified the addition pattern (adding the numbers 1 to 100) embedded within the context; they realized that a way to solve the problem was to calculate the sum of the integers 1 to 100. They then concluded that the total number of pigeons must be greater than 250.

In the course, teachers were asked to make connections between the growth of the dependent variable and the function rule. Some teachers' solutions showed that they made the connection between the growth of the total number of pigeons and the type of function representative of the situation, concluding that because the function was quadratic, the number of pigeons should increase more rapidly than assumed, thereby concluding that 250 pigeons was an unreasonable answer. In some instances, once teachers concluded the function was quadratic, they found the function and used it to find the answer for the hundredth step.

In the course, teachers examined proportionality as a special case of a linear function. They examined multiplicative patterns within data tables of proportional situations and in the function rule. Some teachers solved the pigeon problem by determining if the context represented a proportional situation. Once they realized it was not proportional, teachers concluded that the given strategy assumed proportionality and therefore was inappropriate.

Language played an important role in teachers' mathematics learning, and the final test and journals documented teachers' ability to communicate in writing their new mathematical understandings. Although teachers had many opportunities for communicating their ideas verbally through large-group and small-group discussions, each activity also asked teachers to communicate their ideas in writing. Often, teachers were asked to write sum-

TABLE 15.2

Examples of Responses to the Pigeon Problem and Its Given Solution

Strategy	Sample Response
Estimation (5 responses)	No—the only way to calculate the answer is to add up all the numbers. The student would have had to continue with the table to come up with the correct answer. It is easy to see that this answer 250 is too small because if you start on step 100, where there are 100 pigeons and add a few steps down, $100 + 99 + 98 = 297$. $297 > 250$ and there are 97 more steps of pigeons to add.
Checks out strategy with another example (2 responses)	[*Teacher extends the table to 10 using the pattern that the # of pigeons grows by 2, 3, 4, 5, 6, 7 …; teacher finds that step 5 corresponds to 15 pigeons and step 10 corresponds to 55 pigeons.*] No, if this was true, then the same rationale should work for steps 5 ($15 \times 20 = 300$), 10 ($55 \times 10 = 550$) and the multiplication of these doesn't support his conclusion, because they don't all get the same answer. [*Teacher is testing the use of a scale factor of 20 for step 5 and a scale factor of 10 for step 10*].
Considers type of growth (5 responses)	[*Teacher records the # pigeons grows by 2, 3, 4 and labels the table Quadratic.*] No. The relationship between x and y is a quadratic function. As x increases by 1, y increases by 1 in the second tier. 4 steps = 10 pigeons 100 steps = $25 \times 10 = 250$ pigeons assumes constant growth. With an x^2 in the function the growth is more rapid. [*Shows the work for $x = 1$, $x = 2$, $x = 3$ and $x = 100$ and the function $y = [(x(x + 1)]/2$; finds 5050 pigeons.*]
Makes connection to proportionality (6 responses)	No this is not a reasonable strategy. The student assumed this was a proportional situation, when, in fact it is not. You cannot use a scale factor in non-proportional situations. Solving this with the function, it would look like this: $t = [s^2 + s]/2 = [100^2 + 100]/2 = 5050$ pigeons.

maries of what they learned in their notes; they were directed to translate new conclusions discussed in class into their own words. One teacher commented that until she wrote her journal, she didn't realize how much she really knew. The process of writing, explaining connections among ideas, and documenting their mathematical knowledge informed the teachers themselves about their new mathematical understandings.

FINAL THOUGHTS

After having taught these courses to a large number of teachers, I have found little difference in content knowledge between new and experienced teachers at the start of the course. But teachers do learn mathematics and leave the class feeling better about themselves as learners of mathematics. Elementary school teachers in a supportive environment using a pedagogical model that reflects the *Standards* vision of teaching and learning mathematics can develop deeper understanding of the content they teach and content beyond that level.

As more is learned about elementary school teachers' mathematical understanding and about effective pedagogy for elementary school teachers, more effective professional development opportunities can be developed and implemented. In this article, I identified what I have learned about teaching mathematics to elementary school teachers. But this sharing is not only for those involved with in-service teachers but also those involved in teacher preparation. Initial licensure programs for elementary school teachers have content components to their programs. These courses can be enhanced if those who teach content courses and those who teach methods work cooperatively, sharing our expertise and knowledge about content and pedagogy.

REFERENCES

Lesh, Richard. "Mathematical Learning Disabilities: Considerations for Identification, Diagnosis and Remediation." In *Applied Mathematical Problem Solving,* edited by Richard Lesh, Diane Mierkiewicz, and Mary Grace Kantowski, pp. 111–80. Columbus, Ohio: ERIC/SMEAC, 1979.

Cramer, Kathleen. "Using Models to Build an Understanding of Functions." *Mathematics Teaching in the Middle School* 6 (January 2001): 310–18.

16

Embracing the Complexity of Practice as a Site for Inquiry

Theresa J. Grant

Kate Kline

We noticed also how frequently people sought to reduce the complexity of teaching. Faced with problems, teachers and professional development leaders alike promoted solutions. Students are having trouble learning to subtract? Use manipulatives. Motivation difficult? Teach through games. Children having trouble retaining what they are learning? Increase practice. Grappling with teaching ourselves, we eschewed these flat remedies. We realized that the inherent complexity of practice was a central premise for us. We assumed that the main work of teaching was in the "swamp" of the messy challenges of helping all students learn.

— Deborah L. Ball and Magdalene Lampert,
Issues in Education Research: Problem and Possibilities

IMAGINE that you are about to begin a multiyear project to work with hundreds of teachers in several different districts, all of whom have adopted a new elementary school mathematics curriculum, Investigations in Number, Data, and Space (Investigations). (The Investigations curriculum was developed by TERC with support from the National Science Foundation and in collaboration with Kent State University and the State University of New York at Buffalo.) The majority of teachers were not involved in the adoption decision, and the existing culture for professional development in the districts could be characterized as short-term demonstration by experts. How does one go about supporting teachers in this climate with a new curriculum very different from the traditional ones they had been using? For example, how does one learn to use problem solving and reasoning effectively to enable students to make sense of mathematics? How does one help teachers transition from a traditionally formatted teacher's edition (with the student page surrounded by a few teaching tips) to one that provides a narrative

This paper is based on work supported by the National Science Foundation under grant no. ESI-9819364. The opinions expressed are those of the authors and do not necessarily reflect the views of the National Science Foundation.

description of the lesson with a focus on how to facilitate discussions of the mathematics?

This article describes how the authors designed such a multiyear professional development plan. The Implementing Investigations in Mathematics (InMath) collaborative involved six urban and rural districts that recognized the need for support in their implementation efforts with Investigations and collaborated with the authors to create a professional development program. (InMath was subsequently funded by a National Science Foundation Local Systemic Change grant. For more information on these grants, go to the Web at lsc-net.terc.edu.) We believed it imperative not only to encourage these teachers to think differently about teaching children mathematics but also to encourage them to reconsider what it means to be learners themselves and thus rethink what professional development could be. Using as a guiding force the ideas expressed in the quotation opening this article, we worked to help teachers realize that implementing this curriculum well would require more than simple solutions and quick fixes. We wanted them to acknowledge and appreciate the challenges of teaching and to use their maturing practice as a site for their own learning. Although some sessions focused more on content and others on pedagogy, we strove to keep the complex issues of teaching for understanding central in all sessions.

SESSION DESCRIPTIONS

In this section we describe four types of sessions offered as part of the InMath project.

Content-Focused Sessions

There are a variety of ways one can approach helping teachers develop a deeper understanding of mathematics. In many professional development projects, this is done by introducing teachers to "new" mathematical topics that they may have not encountered, such as discrete mathematics. Instead, we chose to focus on broadening teachers' understanding of the conceptual underpinnings of elementary school mathematics. In this way, we were able to tie the mathematics closely to teachers' practice and embed our work in the daily activities they were already doing with their students. Each year we offered a different series of content-focused sessions centering on the big mathematical ideas in each of the following strands: number and operations, geometry and measurement, and data and change. The sessions were designed around activities from the Investigations curriculum itself, thus providing teachers the opportunity to learn mathematics themselves with understanding. We also organized teachers in heterogeneous, grade-level groups, rather than having a lower elementary and upper elementary school grouping, so that they could more readily reflect on the progression of mathematical ideas throughout grades K–5 and beyond.

Book Walk-Through Sessions

There was no question that, with six to eleven books for each grade level in the Investigations curriculum, "book walk-through" sessions were in great demand. Although they were obviously important, there was a real challenge to structure these sessions in a way that continued to use the complexity of teaching as a site for learning rather than letting them degenerate into basic "how to" sessions. This was accomplished by engaging teachers in doing activities from the units, having them read and discuss the support materials within the books (which provided additional information on the mathematics, the ways students think about the mathematics, and sample conversations), and analyzing assessments within the book to consider how students' thinking develops throughout the unit. Inquiry into the complexity of teaching particular units was also encouraged through "advanced" book walk-throughs for those teachers who had already taught the unit and could then revisit it, thinking in more depth about difficulties in eliciting and enhancing students' thinking developed within the unit. Although one may argue that offering a how-to type of book walk-through session would be worthwhile for someone teaching a unit for the first time, we found it more valuable in the long run to keep a consistent focus on thinking hard about teachers' practice. This had a more wide-ranging impact on the teachers' ability to support students' understanding across units and mathematical topics.

Assessment Sessions

Assessing understanding rather than simply attending to correct answers is a particular challenge for teachers as they shift toward teaching for understanding. The Investigations curriculum uses a variety of types of assessments in order to capture the developing nature of students' understanding. Some of these assessments are embedded in the context of students' everyday work, whereas others are more formal and appear as midunit and end-of-unit assessments. The assessment sessions were structured to allow teachers to analyze particular assessments and anticipate how students might approach the activities and what might challenge them. The teachers were then asked to create criteria to assess students' work and then to use the criteria to analyze students' work samples. Encouraging teachers to think about and create their own criteria rather than providing them with a ready-made rubric is important for deepening their understanding of the mathematics being taught and assessed and for recognizing the value of the information the assessment may provide about their students.

Reflecting on Teaching Sessions

The sessions above were all closely tied to practice in a variety of ways: teachers were experiencing many of the activities in the curriculum with facilitators modeling certain practices and sometimes making these practices explicit through processing discussions. Although the sessions were successful at improving teachers' content knowledge, familiarizing them with the curriculum, and enhancing their understanding of assessment, they did not seem to enable teachers to make the kind of fundamental changes in their teaching that are necessary to implement Investigations well. Additionally, there were many issues around teaching for understanding that warranted intensive consideration and could not be addressed satisfactorily within the existing session structures. Thus we decided to create a professional development format that was not merely connected to teaching but actually used the complexity of practice as the basis for the work of the session.

To allow for focused conversation about classroom practice, the Reflecting on Teaching sessions revolve around videotape of actual lessons from Investigations taught by InMath teachers. One full lesson (typically one hour long) is edited to include a segment on the launch of the lesson, several exchanges among the teacher and students while they are at work, and the closure of the lesson. These segments are then used as a site for reflection during the session. (See Grant, Kline, and Van Zoest 2001 for a more detailed description of Reflecting on Teaching sessions.)

Using an InMath teacher on the tape was an important element of these sessions. It preempted some common reasons for dismissing the results of a videotaped lesson as irrelevant (e.g., the teacher on the tape is an expert; those students are nothing like my students). More important, it helped build a stronger community among the teachers. The teachers who attended the sessions truly appreciated the courage of the teacher on tape and her willingness to open up her practice for analysis. The community was further strengthened by ensuring that the teacher on tape was present at the Reflecting on Teaching sessions. This established a tone of sincere inquiry into *teaching* that did not denigrate into judgments and offhand criticisms of the *teacher* but rather involved collegial discussions of the decisions we make as teachers.

Another primary component of these sessions was to encourage the teachers to think about each phase of the lesson before watching the videotape. For example, before watching the launch, the teachers are asked to read the lesson, consider the mathematical emphasis of the lesson, identify what students might struggle with, and decide whether they think they should deal with any of these struggles in the launch. These "think about" phases bring to the surface some preconceived notions teachers may have about components of a lesson and highlight aspects of teaching that pose challenges for teachers. After watching each videotaped segment, the teachers are encouraged to reflect on what they have seen and discuss the advantages and disadvantages of certain practices.

THE COMPLEXITIES OF LAUNCHING A LESSON

To illustrate the kind of challenges teachers face as they work to teach for understanding, we share here some of the general issues that have surfaced across many sessions during the launch segments. We follow this with a specific example from one of the Reflecting on Teaching sessions. Initial discussions about the launch often revolve around the merits of different approaches to introducing the lesson in question. As the discussion about these approaches progresses, the focus often shifts to determining what it is that students actually need in order to begin the main investigation of the lesson. Answering this question requires one to consider where this lesson is situated, that is, what knowledge and experiences students already have and how students are expected to approach the main investigation of the lesson.

Discussions about how to launch a lesson often spark a lively debate: some believe that students need to be given a certain amount of information or direction up front before beginning the work of the lesson, and others believe that this information should be given individually, on an as-needed basis while students are working. This debate highlights a major philosophical distinction between traditional curricula and standards-based curricula on the issue of how learning occurs—receiving versus constructing knowledge. This philosophical distinction is confounded by a concern for students who will struggle too much and perhaps not be successful. Viewing and discussing the results of a particular launch on videotape have been very powerful in helping teachers understand how giving too much information at the outset can undermine the goals of a lesson and limit the opportunity for students' construction of understanding.

Although some general pedagogical techniques are highlighted by these discussions (e.g., keeping the launch brief, not giving too much away), the majority of the discussion typically centers on content-specific pedagogy. A vignette from one Reflecting on Teaching session is provided below to illustrate this. The fourth-grade lesson for this session occurs late in that year's work on multiplication and division. (Fig. 16.1 provides an outline of the relevant portions of this lesson.) The teacher's guide suggests that the lesson be launched by having students share estimates for 56×4 and 28×15, and "highlight strategies where the students may have multiplied by 10—such as 'I know that 28 times 10 is 280, then there are five more 25's, so that's around 400', or 'Well, 20 times 10 is 200, so it's at least 200'" (Economopoulos, Russell, and Tierney 1998, p. 24).

The main activity of the lesson is for students to find an exact solution for 32×21 using two different strategies, and then represent their solution with

Fourth-Grade Investigations Lesson

Launch:	Estimate 56×4 and 28×15. Share students' estimation strategies.
Exploration:	Find an exact solution for 32×21 using two different strategies.
	Represent solution strategies using equations or words or both.

Fig. 16.1. Lesson overview

equations or words. Although students have not had direct experience with two-digit by two-digit multiplication, they have had a multitude of experiences with multiplication. These experiences include inventing procedures for solving two-digit by one-digit problems and significant exposure to the structure of numbers through analyzing patterns and breaking apart numbers into manageable chunks (e.g., 24 can be broken up in a variety of ways, including $20 + 4$, 12×2, $25 - 1$, and so on). This particular activity, solving 32×21, is intended as "one way of assessing students' understanding of multiplication in general, and the kinds of connections that they are making among related multiplication problems" (Economopoulos, Russell, and Tierney 1998, p. 25).

After coming to consensus on the main mathematical goals of this lesson and discussing some of the mathematical ideas that students might struggle with during the lesson, the discussion in the Reflecting on Teaching session shifted to whether or not any of these struggles should be addressed in the launch. The discussion that follows is one portion of a ten-minute discussion on whether or not additional estimation problem(s) should be added to the launch. Although parts of the discussion have been omitted for clarity, what remains provides the essence of the type of discussion the teachers were having about this aspect of the launch. (In this vignette, F indicates the facilitator of the session; T1, T2, and so on indicate individual teachers participating in the session.)

T1: I think after you estimate 56×4, there wouldn't be any harm in just reviewing with them.... "How would you break this number apart to solve this particular problem?" Maybe do that not once, but maybe do it with several other problems, just so that they do know that they have an understanding of how to break a number apart to do multiplication.

F: Are you talking about estimating it or solving it?

T1: I'm saying in the launch, after you've done your estimating.... After you've done that, and you're ready to go into "We're going to now do a problem," and you put on the board 28 × 15. That might be kind of frightening to them, in a sense, to come up with a strategy for how to break it apart.... So I would have to say to them, "You know what, before we do a problem like this, let's kind of talk about some strategies that we used when the numbers were smaller."

T2: Not necessarily solve it, but ...

T1: Exactly. What strategies did we do? Like when we had a number like 25 and we multiplied times something. What did we do to solve that problem when we were doing it before?

T3: If I got a bunch of blank stares, I'm probably going to take them back ... "We've just had that discussion [about estimating 56 × 4]. Now how can we use those ideas to help us estimate this new problem [28 × 15], rather than give them another new problem.

F: Okay, what do the rest of you think about this—needing to think about breaking apart 28 × 15 or reviewing breaking apart?

T4: I don't think that after you've done that [estimated 56 × 4], that it would hurt them to struggle a little bit on this, on the next one [28 × 15]. Like, instead of talking to them about strategies and what we've done in the past—that would be part of my lesson later on. You know what I mean? And if we're going to talk about launch, I would just launch it, you know, and bring up the 56 × 4 and what we did with that, but let them struggle a little bit with estimating this one [28 × 15]. And then go into a lesson later and talk about what happened: "What did you do? What worked/what didn't work?"

In this episode, the teachers generally seemed comfortable with their students' abilities to estimate 56 × 4; some, however, were expressing concern that students would struggle with estimating 28 × 15. Interestingly, the fact that teachers may agree on the identification of a particular struggle did not imply that they agreed on what should be done about it. The first teacher expressed the need to have a discussion about breaking numbers apart and perhaps do several problems with the students to get them comfortable with this before moving on to 28 × 15. The next teacher suggested that simply discussing the strategies used in estimating 56 × 4 would be a sufficient review, whereas the last teacher in this episode proposed that letting the students struggle with estimating 28 × 15 was fine. At this point the facilitator invited other teachers to get involved in this debate.

F: Okay, well, what about the rest of you? What are you thinking about this? This is an important issue. How do you think they're going to attack those two estimations? How much are they going to struggle, and how much is reasonable for them to struggle?

T5: We were talking about the fact that from 56 × 4 to 28 × 15 is a big jump for a lot of kids. So we're wondering, couldn't you do 56 × 4, and then for the second estimation problem, keep the 56 and change the 4 to a 14? So that they're keeping their same 56 framework there, and they're keeping the 4, but they're adding the 10 on to that.

F: What do you think?

T5: That might be a little comfort point, as a bridge between the two, before you jump to 28 × 15 and have two brand new numbers.

T6: That's what we were just saying. It seems like such a jump, because [estimating] 56 × 4 you can use the landmark [50], but then way over here [is 28 × 15]. So, it's a huge jump.

[Several teachers are talking over one another.]

F: Any counterpoints to this from those of you who have been silent? Yes, what's your counterpoint?

T7: I guess by adding on a 10 and making it 14, to me that kind of sends a submessage that we are still pushing them to get the right answer, rather than looking at a different number.

F: When you say the right answer, you mean …

T7: An exact answer.

F: Oh, an exact answer.

T7: Yeah, I think we need to just look at two separate numbers [problems] like they do here and push it into the tens, and just add on to the ideas like (T4) said in the beginning. By just putting a 1 on there, I think we're still kind of pushing them into [thinking] we want an exact answer.

T8: We've done that in "Landmarks in the Thousands" (previous unit in curriculum). *[Other teachers talking.]* No, they're not all going to have it, but at some point you

> have to stop and you have to keep going, and hope ... you can't drive it and drive it and drive it and drive it home, so that everybody gets it. At some point you just have to launch it.

In this segment of the conversation another modification to the launch is suggested: first estimate 56×4, then 56×14 (before attempting 28×15). The teacher who provided this suggestion, and those who murmured in agreement, seemed to be attempting to construct a sequence of problems that would allow students to incorporate one small change into their strategy in order to solve the new problem. However, it is this very nature of the suggestion that led one teacher to conclude that this may have the effect of focusing on finding exact answers rather than estimates. It led another teacher to remind the group that students had already had enough experience moving from multiplying by 4 to multiplying by 14 in a previous unit.

The discussion about whether or not to supplement the launch of the lesson with additional estimation problems highlights a variety of issues involving both mathematical content and how students come to learn. With respect to the mathematical content, there seemed to be those who did not fully understand the purpose of having students estimate (both in general and with respect to this lesson) or what knowledge a student would need in order to get an estimate. The discussion of the main problem in this lesson—finding two ways to solve 32×21—requires that students use related multiplication problems to solve this problem. Estimates of 32×21 (like 32×20) are often ideal starting places for reaching an exact answer. Another issue that may have clouded the role of estimation in this lesson was the fact that 56×4 was not just easy to estimate ($50 \times 4 = 200$) but was relatively easy for many of their students to go ahead and solve exactly (by adding 6×4 to the estimate of 200). Thus some teachers saw this as a nice opportunity not only to practice estimation but to find an exact solution as well. An unfortunate result was that this shifts the focus of the launch to finding exact answers, which is problematic when considering 28×15.

It is important to note that these teachers were thinking hard about how to launch this one particular lesson, and that this kind of discussion elicited an analysis of the particular mathematical ideas students were working on in this lesson (including evaluating the choice of numbers used in the problems). However, in order to regain perspective, one needs to situate this lesson in the larger scheme of how these ideas are being developed throughout the year-long curriculum. The final teacher in the vignette above began to raise this issue by reminding the group about the experiences students have already had in a previous unit that prepares them for this lesson. As the discussion continued, another teacher pointed out that this was the first time that students were solving a two-digit by two-digit multiplication problem, and that it was intended to be an opportunity for teachers to see how students extended their understanding of multiplication to a more complicated problem. This vacillation between the specifics of the activities in a lesson and the larger context of the yearlong study of mathematics is a necessary aspect of dealing with the content and content-specific pedagogical decisions that teachers make.

Perhaps the most important thing that this example highlights is the honest inquiry into the teaching and learning of mathematics elicited in these sessions. These teachers were thinking hard about the "messy" challenges of helping students learn about multidigit multiplication. They provided arguments for what they thought should and should not be done in the launch of the lesson and gave rationales for those arguments. They listened to one another's points of view and were willing to offer counterarguments or concurrence. There are no simple answers to the question of how one should proceed with a particular lesson or with a particular group of students. However, this kind of inquiry into the complexity of teaching can affect teachers in powerful ways. It brings to a conscious level some notions about teaching and learning that need to be considered from different perspectives so that teachers will be better able to make decisions when preparing to teach, when teaching a lesson, and when reflecting on that lesson.

This example also highlights the nature of the facilitator's role in establishing the tone necessary to build and support a community based on reflective inquiry into practice. In this session, the facilitator was not viewed as an "expert," dispensing knowledge about how one should launch the lesson. Instead, she attempted to foster discussion of the advantages and disadvantages of the suggestions being made and to promote an analysis of the impact these suggestions have on students' learning. The facilitator also took responsibility for summarizing the various ideas that had arisen in the discussion and gave teachers the opportunity to consolidate them. In the session described above, the facilitator concluded that portion of the discussion by placing the lesson in context (with respect to what students had learned in previous units) and by acknowledging the disagreement about how much support students needed in the launch. Finally, she asked the teachers to consider the fact that this lesson included an assessment and that it was therefore designed to provide teachers with information on how students were thinking. In this way, the facilitator refocused the discussion on what teachers might gain from observing how students approach their first two-digit by two-digit multiplication, and thus on the potential advantages for letting students struggle with this problem.

MOVING BEYOND THE LAUNCH

Similar discussions ensue during the remainder of the daylong Reflecting on Teaching sessions as the focus shifts to the challenges teachers face as they interact with students at work and finally as they think about how to close the lesson. For example, in thinking about interacting with students as they work, teachers are challenged by how best to support students who seem to be struggling without simply telling them how to get through the activity at hand. They recognize that the latter has short-term benefits but are concerned about the long-term repercussions. We have also had some debate about what it means to develop students who are "independent" problem solvers and the actions necessary from teachers to encourage this characteristic in students. It takes a great deal of care, consistency, and vigilance to establish and maintain norms in the classroom where students are empowered to do their own thinking rather than waiting for assistance from the teacher. Watching teachers on tape who have attempted to establish these norms has prompted further reflection on ways in which this can be accomplished.

Thinking about how to close a lesson has also highlighted primary issues that challenge teachers when teaching for understanding. Although many have lamented the fact that they often run out of time for a significant closure, they have also identified several different purposes that make the closure crucial. One purpose for a closure may be to provide students the much-needed opportunity to consolidate their ideas about the mathematics they have been doing. Another possibility may be to discuss a particular misunderstanding or incomplete understanding to discern what students know and help them realize that they will continue to investigate and develop these ideas. A related issue is how decisions are made about which ideas from students are shared and when. Take, for example, a strategy that only a couple of students have used and explored. If it is particularly sophisticated, the teacher must decide whether to allow these students to introduce it to the rest of the class and consider whether the other students will be able to understand the strategy. These are complex decisions that have a profound impact on students' learning. By reflecting on videotaped closures where different approaches are illustrated, the teachers are able to consider the impact of these decisions on students.

CONCLUSION

As teachers work to implement new mathematics curricula, they are challenged to rethink their ideas about the nature of mathematics, to learn new mathematics content, and to enact a pedagogical approach that supports students' understanding. This requires a substantial shift in their thinking and a conscious monitoring of ways of interacting with students. Enabling such a shift demands that we carefully design professional development programs to balance the competing, yet complementary goals of helping teachers focus on individual aspects of teaching while realizing that teaching is too complex for quick fixes or easy solutions. We must provide opportunities for teachers to embrace the complexity of teaching as a site for inquiry and build communities whose members are willing to think together about their practice. By offering these multiple opportunities over an extended period of time, we allow teachers to develop more grounded, general notions of teaching and learning mathematics with understanding.

References

Economopoulos, Karen, Susan Jo Russell, and Cornelia Tierney. *Packages and Groups.* Palo Alto, Calif.: Dale Seymour Publications, 1998.

Grant, Theresa J., Kate Kline, and Laura Van Zoest. "Supporting Teacher Change: Professional Development That Promotes Thoughtful and Deliberate Reflection on Teaching." *NCSM Journal of Mathematics Education Leadership* 5 (fall 2001): 29–37.

17

The Reflective Teaching Model
A Professional Development Model for In-Service Mathematics Teachers

Lynn C. Hart

Deborah Najee-ullah

Karen Schultz

IN RESPONSE to research on learning, the last 20 years has seen significant change in the way we think about grades K–12 mathematics education. Mathematics educators have not, however, limited their thinking about change to grades K–12 classrooms. We have had to rethink the way we deliver instruction to both students and teachers. This raises many questions about how to create significant professional development experiences for practicing teachers. Can we apply what we know about how students learn to work with grades K–12 teachers? Should we balance the learning of mathematical content with the learning of how to teach from a constructivist perspective? Do teachers need to be aware of the beliefs they hold about teaching and learning? Will knowledge about their beliefs affect their instructional decisions? Does regular reflection on their learning precipitate greater change?

As developers of the Reflective Teaching Model (RTM), a professional development model for in-service mathematics teachers, we sought to answer these and other questions when we first implemented the RTM in the Atlanta Math Project (Grouws and Schultz 1996) nearly a decade ago. Our goal was to develop experiences that challenged traditional, existing frameworks and offered alternative strategies for teaching. First, we had to explore what we believed about teaching and learning. We had to develop experiences that gave teachers insights into their own teaching and into how children learn. We had to try out our ideas and revise our work as we moved forward. In short, we became learners ourselves. The publication of the *Curriculum and Evaluation Standards for School Mathematics* (National Council of Teachers of Mathematics [NCTM] 1989) offered the motivation for that first implementation of the RTM and provided a guidepost for our work.

Today the NCTM *Principles and Standards for School Mathematics* (2000) and the National Board Professional Teaching Standards (2003) affirm the value of the RTM as a professional development tool.

Framework

The RTM is grounded in the theories of *constructivism* and *metacognition.* It is also based on our assumptions about the value of *modeling, sharing authority, reflecting,* and *heuristic teaching.* Collectively these constructs guide the development of all activities and experiences in the RTM.

Constructivism is a theory of how people learn (von Glasersfeld 1983). From a constructivist perspective, learning is an active process of constructing knowledge rather than the passive reception of information.

Metacognition is a theory about how we think. The theory of metacognition (Flavell 1979) refers to our ability to think about our thinking, to monitor and regulate what we are doing and thinking while we are having an experience (for example, solving a problem) so we can gain more from the experience.

Modeling demonstrates for teachers the environment we are trying to construct. It addresses the problem that "Teachers seeking to change their practice may not have useful images from their personal experience to guide the creation of a focused and productive classroom culture that emphasizes inquiry and exchange of ideas" (Goldsmith and Schifter 1997, p. 25).

Sharing authority (Cooney 1993; Hart 1993) means that all participants in the RTM, whether teachers, teacher educators, or students are seen as learners, and all are seen as teachers. It builds trust, ownership, and cohesion among those involved.

Professional growth is stimulated when teachers engage in experiences that challenge them to think deeply about their beliefs, practices, and assumptions about teaching and learning. This is promoted through reflection, most often prompted by a peer, while planning lessons and debriefing after.

Teaching heuristically (Rachlin, Matsumoto, and Wada 1985) requires viewing teaching as a problem to be approached and solved in much the same way a mathematics problem can be approached and solved—heuristically, as opposed to algorithmically. From this perspective there is no script; rather, the lesson plan becomes a set of strategies that may be used for the teaching solution, a solution that will likely be modified as needed during the lesson.

Format

Our model for change follows a model/experience/reflect format (see fig. 17.1). The facilitators first model an exercise such as planning, teaching, or problem solving; later, the teachers experience the activity; and at the close of each activity, everyone reflects. As an activity in the RTM cycles through the model/experience/reflect steps, assumptions of the RTM are realized. Teachers construct new knowledge about teaching and learning (constructivism). They learn to monitor their thinking and behavior (metacognition). They expand and enhance their frameworks for teaching by watching how others think about and teach from a reform perspective (modeling). They have a learning experience where their ideas are valued; they experience teaching where they value the ideas of their students; and, they experience planning collaboratively with teacher educators (sharing authority). They develop strategies for lesson planning that assist them in "solving" the teaching problem (heuristic teaching). The format and framework of the RTM are pictured in figure 17.1.

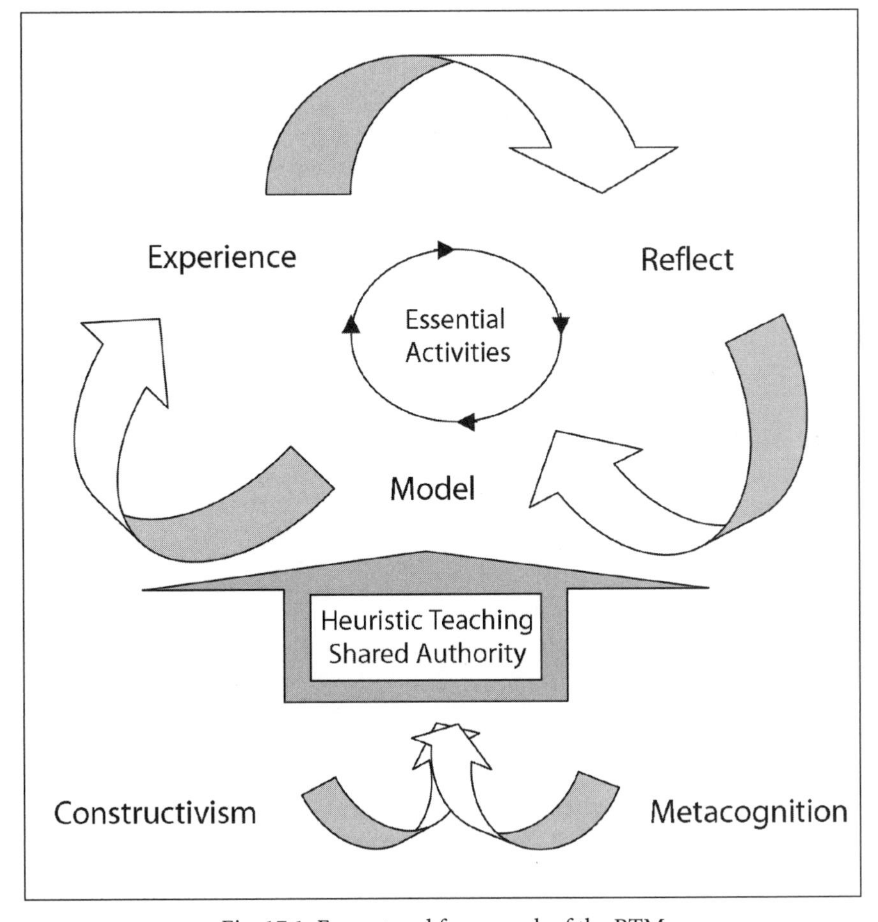

Fig. 17.1. Format and framework of the RTM

ESSENTIAL ACTIVITIES IN THE REFLECTIVE TEACHING MODEL

Each project in which we implement the RTM is facilitated by two mathematics educators. Within the model/experience/reflect conceptual framework teachers engage in several essential activities. Each of these will be described in some detail, since they make up the "content" of the model.

Initial In-Service Program

All projects begin with an in-service program of five to ten days that introduces the philosophy and language of reform, lays the groundwork for sharing authority, and builds the relationships necessary to carry on a long-term partnership for support during the school year. The in-service program consists of a set of carefully developed activities that allows teachers to experience an idea or concept as an observer and then reflect on the activity. The teachers then experience the activity firsthand and reflect on it: thus the model/experience/reflect format. Components of the in-service program include Thinker/Doer problem solving, the Plan/Teach/Debrief sequence, and various forms of reflection. Most projects also have a grade-level focus, a mathematical content focus, and a pedagogical focus.

Thinker/Doer

Thinker/Doer is a paired, problem-solving activity adapted from Whimbley and Lochhead (1986). Through this activity much of the mathematics content of an RTM project is developed. Teachers engage in systematic problem solving through the Thinker/Doer activity that uses problems selected to align with the content and grade-level focus of the in-service program.

In the activity, the Thinker functions as a coach or facilitator, listening as the Doer solves a problem. The Thinker asks questions that facilitate thinking without providing direct instruction or judgment, prompting the Doer to communicate her mathematical understandings to the Thinker. To do this, the Doer must monitor her thinking as she works on a problem. The two teacher educators first model the process for the in-service teachers. One plays the Thinker, and the other, the Doer. The Thinker gives the Doer a problem the Doer has never seen before. The Thinker asks the Doer to consider the following questions:

- Do you think you understand the problem?
- Do you think this problem is hard or easy?
- What strategies do you think you will use?
- How do you think you will do? Why?

After responding to the questions, the Doer thinks out loud as she solves the problem. The Thinker can only ask clarifying questions. This is crucial. The Thinker is imitating the behavior of the classroom teacher as a coach or facilitator and should not be "teaching" the problem. After the Doer has solved the problem or when the Doer is stumped, the Thinker asks the same four questions of the Doer.

First, the in-service teachers are given the problem to solve and discuss, so that while the teacher educators model the Thinker/Doer activity they can concentrate on the process. After the Doer teacher educator responds to the last set of questions, all the participants discuss the mathematics of the problem and additional processes for solving it. Finally, the two teacher educators reflect on the Thinker/Doer experience in front of the teachers, after which the teachers are invited to join in the discussion. We discuss how the activity went, and we reflect on the purpose of the two roles. A teacher educator solving a problem that is new to her in front of the teachers can serve as a powerful way to humanize mathematical problem solving and show the struggle most people go through when confronting a new problem. It almost instantly levels the playing field and builds collegial relationships.

On the next meeting day, the teachers experience doing a Thinker/Doer activity themselves. The class is divided into two groups, and each group discusses a problem they have worked on prior to coming to the workshop. The group meeting is important so that all the Thinkers have already solved the problem and can listen attentively to their partner. After the two separate group discussions, the partners meet in their Thinker/Doer teams. Teachers keep the same partner for the entire in-service program, meeting daily. Each partner takes her turn at solving a problem cold (being the Doer), and each partner plays the Thinker and listens to her partner solve a problem. They struggle with the mathematics, with how to communicate their mathematical understandings, and with the teacher's role in a problem-solving situation. At the conclusion of each exchange, the group reassembles and talks about the content that was explored and about the difficulty in each of the roles. The Thinker/Doer activity is done daily with new problems. Teachers often comment at the end of several days of Thinker/Doer that it is hard to solve a problem without their Thinker. In addition to building a mental model of the teachers' role when students are solving problems, this activity enhances teachers' metacognitive ability. The Thinker asks the kinds of questions that a problem solver should be asking herself while solving a problem. Having those questions asked repeatedly encourages a person to ask them of herself or her students. For a detailed discussion of Thinker/Doer, see the Sixty-sixth Yearbook Professional Development Guidebook, Chapter 5.

The Plan/Teach/Debrief Sequence

The Plan/Teach/Debrief sequence is a form of cognitive coaching. During the initial in-service meetings, the teachers see the Plan/Teach/Debrief sequence modeled. During the school year follow-up, they engage in monthly Plan/Teach/Debrief sequences with their teacher-educator partner. In subsequent years, teacher pairs work together to think aloud through the process of planning a lesson, teaching the lesson, and debriefing on the lesson. In order to make the sequence "work" in the initial in-service meetings, the actual planning for a mathematics lesson is videotaped ahead of time. One of the teacher educators teaches the lesson to the teachers. The group reflects on the experience and then views the planning video. In the debriefing, the teacher educator who taught the lesson discusses how the lesson differed from her original plans, what worked, what didn't work, and what she would have done differently. The following set of general questions guides the debriefing both in the modeling and during the school year.

- How do you think the lesson went? [Always ask this question first.]
- What do you think worked well? [probe: Do you think anything in particular supported students' learning?]
- What do you think didn't work well? [Probe for examples of interference with students' learning.]
- Can you think of particular examples that made you think students were learning (or not)? What could you have done differently if they were not?
- What overall would you do differently?
- Where will you go from here in the lesson?

The first question is particularly crucial. As teachers are "changing" their practice, the "change" may not be evident to the observer. It is important to get the perspective of the person who actually taught the lesson at the beginning of the debriefing and to use their comments to guide the debriefing. Also, the questions listed after the initial question try to focus on helping the teacher become more aware of what he or she is doing in the classroom that is facilitating students' learning and what he or she is doing that may be interfering with students' learning. We help the teacher look for evidence from the students' comments and actions. The debriefing is not an evaluation of the teacher. It is a method for assisting the teacher in learning how to think critically about his or her teaching. Every attempt is made to follow the lead of the teacher and to be sensitive to his or her concerns. Insensitivity at this point could ruin further collaboration.

Mathematics content for the lesson is chosen carefully to support the mathematics focus of the in-service program and to align with the grade level of the teachers. The teachers engage in one Plan/Teach/Debrief sequence with one another during the initial in-service program and then monthly with a teacher educator or colleague in their own classrooms.

Forms of Reflection

During the initial in-service program the teachers engage in several forms of reflection. They complete a reflection log at the end of each Thinker/Doer sequence. They reflect orally throughout the day after each activity. At the end of each day, the two teacher educators put on their teacher hats and reflect publicly on the day, sharing their concerns, issues, and successes of the day. After the teacher educators model the process, the teachers are invited to join the conversation. This process develops slowly over the initial in-service program, but by the end most teachers are much more willing to share openly their concerns and what they have enjoyed about their work that day. This reflection is used to assist planning for the next day.

Videotaping is first introduced in the initial in-service program as a form of reflection on teaching. At the beginning, only the teacher educators are videotaped, and the tapes are used to facilitate reflection. During the school year, teachers begin to videotape themselves. Videotapes provide a candid visual record of rich verbal and nonverbal dynamics of classroom episodes. A videotape of yourself can make you aware of the things that you do that are second nature, but may or may not be productive in the classroom. Videotaping can help teachers identify what is working, what is not, who is doing the talking, and to whom they are talking. It provides a record of the discourse that can be used to make pedagogical decisions about future instruction.

Content, Process, and Pedagogical Focus

The mathematical content focus for an RTM project is dictated by the grade levels of the participating teachers and the needs and requests of local school systems. Problems for Thinker/Doer and the Plan/Teach/Debrief sequence are selected to support the content focus.

Problem solving is the main process on which the RTM focuses, although communication, reasoning, connections, and representations all play roles. In particular, we focus on three problem-solving competencies identified by Krutetskii (1969) as crucial in the problem-solving process: flexibility, generalizability, and reversibility. For example, if we were conducting a workshop for elementary school teachers and our content focus were whole-number operations, the following might be used at various times for a small group or whole-class problem-solving sessions.

(1) Flexibility refers to the ability to solve a problem in more than one way. *Find at least three ways to find the product for 25×36.*

(2) Generalizability refers to the ability to see patterns, draw conclusions, and deduce a rule.

Study the following addition problems. What do you notice?
25 + 10   26 + 9   27 + 8   28 + 7   29 + 6 (and so on)

(3) Reversibility refers to the ability to switch directions in solving a problem.

Make up a division problem that has a quotient of 23 and a remainder of 5.

Our focus for developing pedagogical content knowledge is aligned closely with the NCTM *Professional Standards for Teaching Mathematics* (1991). We try to help teachers develop a model of their role and the role of the student in discourse. A model of the teacher as coach is presented through classroom modeling as well as through videotaped segments of other teachers. Current resources on the NCTM Web site, www.nctm.org, provide many excellent examples of classroom scenarios for reflection as the Annenberg/CPB Math and Science Collection does (Annenberg/CPB 1997a, 1997b).

School Year Follow-Up

All teachers in the RTM engage in monthly Plan/Teach/Debrief sequences during the school year. This is done in the first year with the teacher educator coaching a participating teacher. In subsequent years of multiyear projects, teachers are developed as mentors to coach other teachers in their own school.

Initially, the coaching involves modeling the self-assessment and self-management of the thinking engaged in by the teacher educator as she plans, teaches, and reflects on a mathematics lesson. Coaching from this perspective is based on our understanding that metacognition (being aware of one's thinking processes) fosters independence in learning by providing personal insights into one's own thinking. It builds the problem-solving skills necessary to teach from a reform perspective. Through the year, as the teacher educator and teacher continue the Plan/Teach/Debrief sessions, the teacher educator does less modeling, encouraging the teacher to reflect on her own thinking.

The collaboration occurs over a two-day period. On the first day the teacher educator and classroom teacher plan a lesson together. The next day the lesson is taught and debriefed. The first lesson is modeled by the teacher educator, allowing the classroom teacher to observe her students in a reform environment. Over time the teachers become comfortable with the partnership and responsibility for teaching is turned over to the classroom teacher. However, the lesson is always seen as a shared product to which both partners contribute. Lessons are videotaped and used privately for reflection and, on occasion, publicly for future in-service activities to show examples of the struggle. As teachers move into mentoring roles in their second year, a Plan/Teach/Debrief Guide (Hart and Najee-ullah 1997) is used to assist teachers in the transition from student to coach.

The full group meets regularly during the school year to share "how it is going" with their colleagues, to raise questions and concerns about the process, and to engage in further explorations of content and the *Standards.* These meetings serve to continue the educational process, and to provide opportunities to develop and strengthen collegial ties with other teachers. (See fig. 17.2 for a summary of the RTM essentials.)

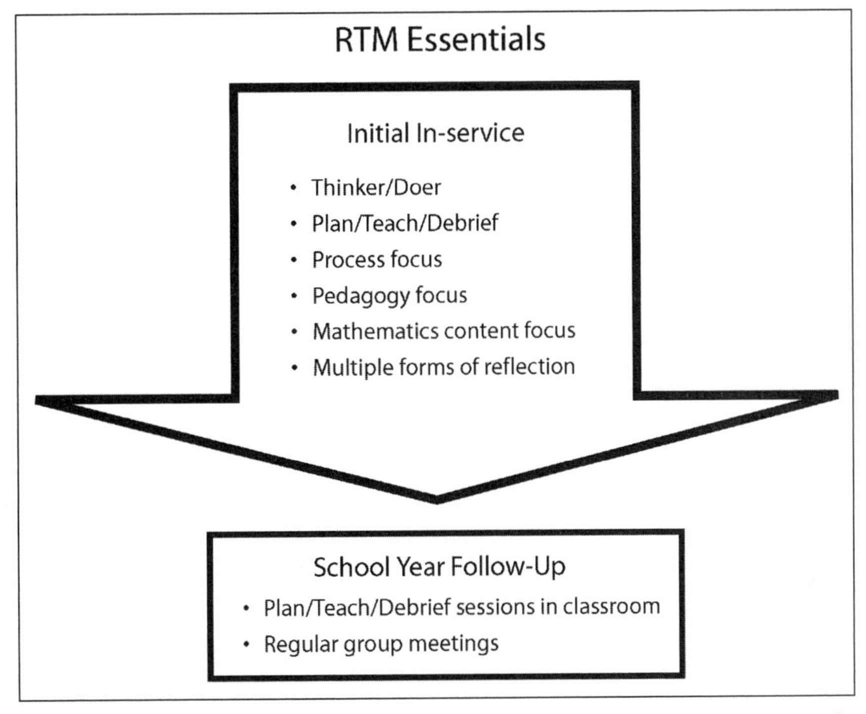

Fig. 17.2. Essential activities in the RTM during the initial in-service program and during the school year follow-up

RESEARCHING THE RTM

In our work we have collected many documents and vignettes from participants about their experiences in the RTM. What follows is an excerpt provided by a teacher who participated in the Atlanta Math Project, the first project that implemented the RTM. The excerpt comes from midyear in a seventh-grade classroom. The teacher had been attempting to change her pedagogy to support a constructivist theory of learning. Many of the students had "bought into" the classroom, appropriate norms and values, but several of her inclusion students were still hesitant to share their strategies or thinking with the class. The excerpt is subtle, yet to those who have attempted the process it provides a picture of a very powerful moment in a teacher's life as she came to understand the value of reform through the eyes of her students.

> I was teaching 7th grade and had an inclusion class. The inclusion teacher lined up her class of seven young men along the back row. They sat silently until December, when one young man finally raised his hand to come to the overhead and show how he worked [a] problem. The problem was a simple addition of fractions problem. Well, this young man said that he couldn't do fractions, but he liked adding decimals. It took forever and the class sat silent and mesmerized as he changed each fraction to a decimal, added the decimal numbers, then changed the answer back to a fraction. The class was in awe. They had never thought about the problem that way before. They thought that his way was so neat and many understood, for the first time, addition of fractions!
>
> It wasn't the traditional way or the quickest way to address the problem and had I stopped to show him the "right" way, he would have silently watched me, returned to his seat, and never said another word that year. But as we watched and we listened to him think through the mathematics in a way that made sense to him, I truly understood what the [RTM] was attempting to accomplish.

The excerpt above describes a specific RTM experience that fostered an "aha" moment and, gradually, growth and change in one teacher. It illuminates the power of students' thinking and how a teacher can benefit when she reflects on experiences such as the one just described.

From accounts similar to the one above written by teachers and administrators who participated in projects with the RTM, from interviews, and from observations, we have learned that RTM participants experience change in many ways. They change in their (1) pedagogical content knowledge, (2) content knowledge, (3) teaching behaviors, (4) awareness of students' processes and learning, (5) professional involvement and advancement, and (6) attitudes and beliefs about teaching and learning mathematics.

FINAL COMMENTS

Frequently, when teachers, administrators, or school districts learn about an innovation in mathematics education, they implement it for a while but not long enough for the benefits to manifest themselves. They abandon the practice, concluding it did not work. Implementing the RTM is a long-term commitment by teachers and administrators. This kind of commitment promotes reflective practice and improvements in children's mathematics learning.

The elements that have made the RTM successful are difficult to isolate. However, we will attempt to discuss briefly two of the crucial elements. First, the theoretical assumptions that provide the foundation and framework for the RTM have been operationalized. We have carefully crafted activities to engage teachers in experiences that nurture behaviors aligned with our underlying beliefs. The process of transforming these beliefs into practice begins as the beliefs are raised, discussed, and sometimes challenged in the context of RTM activities. It continues implicitly when each activity is modeled by the teacher educators, experienced by the teachers, and then processed during the reflections. The process of transformation continues as this recursive cycle is repeated for every component of the RTM.

Second, it is crucial that there is a community of learners that develops as the teachers and teacher educators support each other in the process of change. This supportive community is grounded in the willingness of teacher educators to share authority and become partners with the teachers involved. Every activity the teachers engage in, the teacher educators do first, to model, but also to establish them as willing partners in the process. This partnership entails every component of the RTM activities. From Thinker/Doer problem solving to planning, teaching, or reflecting on their lessons, the teacher educator becomes a colleague and a partner in every element of the change process.

REFERENCES

Annenberg Foundation/CPB. *Teaching Math: A Video Library, K–4.* Produced and directed by the Science Media Group of the Harvard Smithsonian Center for Astrophysics. Videocassettes. Washington, D.C.: Annenberg Foundation/CPB, 1997a.

———. *Teaching Math: A Video Library, 5–8.* Produced and directed by the Science Media Group of the Harvard Smithsonian Center for Astrophysics. Videocassettes. Washington, D.C.: Annenberg Foundation/CPB, 1997b.

Cooney, Thomas. "On the Notion of Authority Applied to Teacher Education." In *Proceedings of the 15th Annual Meeting of the North American Chapter of the Psychology of Mathematics Education,* edited by Joanne Rossi Becker and Barbara Pence, pp. 40–46, Vol. 1. San Jose, Calif.: San Jose State University, 1993.

Flavell, John. "Metacognition and Cognitive Monitoring." *American Psychologist* 34, no. 10 (1979): 906–11.

Goldsmith, Lynn T., and Deborah Schifter. "Understanding Teachers in Transition: Characteristics of a Model for the Development of Mathematics Teaching." In *Mathematics Teachers in Transition,* edited by Elizabeth Fennema and Barbara Scott Nelson, pp. 19–54, Mahwah, N.J.: Lawrence Erlbaum Associates, 1997.

Grouws, Douglas, and Karen Schultz. "Mathematics Teacher Education." In *Handbook of Research on Teacher Education,* edited by John Sikula, Thomas J. Buttery, and Edith Guyton, pp. 442–58. New York: Simon & Schuster Macmillan, 1996.

Hart, Lynn C. "Shared Authority: Roadblock to Teacher Change?" In *Proceedings of the Fifteenth Annual Meeting of the North American Chapter of the Psychology of Mathematics Education,* Vol. 2, edited by Joanne Rossi Becker and Barbara Pence, pp. 189–97. San Jose, Calif.: San Jose State University, 1993.

Hart, Lynn, and Deborah Najee-ullah. "The Plan/Teach/Debrief Guide for K–12 Teachers." Georgia State University, 1997. education.gsu.edu/lhart/ as of September 5, 2003.

Krutetskii, Vadim A. "An Investigation of the Mathematical Abilities in School Children." In *The Structure of Mathematical Abilities,* Vol. 2, Soviet Studies in the Psychology of Learning and Teaching Mathematics, edited by Jeremy Kilpatrick and Izaak Wirszup, pp. 5–58. Chicago: University of Chicago, 1969.

National Board for Professional Teaching Standards. Available at www.nbpts.org/about/coreprops.cfm/, 2003.

National Council of Teachers of Mathematics (NCTM). *Curriculum and Evaluation Standards for School Mathematics.* Reston, Va.: NCTM, 1989.

———. *Professional Standards for Teaching Mathematics.* Reston, Va.: NCTM, 1991.

———. *Principles and Standards for School Mathematics.* Reston, Va.: NCTM, 2000.

Rachlin, Sidney, Annette Matsumoto, and Li Ann Wada. "Teaching Problem Solving within the Algebra Curriculum." Paper presented at the annual meeting of the American Educational Research Association, Chicago, April 1985.

von Glasersfeld, Ernst. "Learning as a Constructive Activity." In *Proceedings of the Fifth Annual Meeting of the International Group for the Psychology of Mathematics Education,* Vol. 1, edited by Jacques C. Bergeron and Nicholas Herscovics, pp. 41–69. Montreal: University of Montreal, October 1983.

Whimbley, Arthur, and Jack Lochhead. *Problem Solving and Comprehension,* 4th ed. Philadelphia: Franklin Institute Press, 1986.

18

Encouraging Professional Growth and Mathematics Reform through Collegial Interaction

P. Mark Taylor

WHAT supports at the building or district level can enhance the growth of mathematics teachers? Structural supports and leadership that encourage collegial interaction—that is, collaboration among colleagues that is focused on common goals—hold that potential. Collegial interaction can bring about professional growth for the individual mathematics teachers while raising the achievement levels of their students. Extensive reform requires this kind of team effort, and teams are more likely to be successful if team members are truly colleagues. This article addresses some basic issues related to the impact of collegial interactions, identifies which interactions should be promoted, and describes how to support collegial interaction.

Why Focus on Collegial Interactions?

The National Council of Teachers of Mathematics (NCTM), through its *Standards* documents (NCTM 1989, 1991, 1995, 2000), has set the stage for sweeping changes in teachers' ideas about what mathematics should be taught, how it should be taught, and how it should be assessed (Lappan 1999). The importance of the role of collegial interaction was identified early in the process in the depiction of the vision of how such changes should be accomplished (NCTM 1989, p. 253, emphasis added):

> Although we are confident that many teachers are now ready to teach the kind of mathematics program outlined in the *Standards,* many others will need and demand additional training or refresher courses. These programs must be developed in collaboration with the teachers. *They must include mechanisms for sustained collegial interaction,* links between staff development and classroom practice, and the participation of administrators to ensure support for the proposed changes.

This statement was followed by a description of the roles that school and district administrators need to play, including the responsibility of "supporting teachers in self-evaluation and in analyzing, evaluating, and improving their teaching with colleagues and supervisors" (NCTM 1991, p. 181).

Stigler and Hiebert (1999) noted that "a requirement for beginning the change process is finding time during the workweek for teachers to collaborate" (p. 144). This conclusion was based partially on the Japanese mechanism for steady improvement of instruction called the *lesson study*. The design and sharing of collaborative, research-based lessons has made them an integral part of ongoing improvement in practice, not only for the individuals involved but also for the broader community of Japanese mathematics teachers.

Although lesson study is not currently part of the U.S. school culture, there have been efforts to use collegial interaction as a platform for reform in American schools (Friel and Bright 1997). One emerging format for this kind of reform is that of a professional development team. A team consists primarily of the faculty of a mathematics department at a middle school or high school, one or more student-teachers, and a university faculty member. Together, they embark on an ongoing cycle of choosing a goal or problem, gathering data, collecting ideas about what and how to change, implementing their plan, and assessing the results. The next section of this article addresses the specific types of interactions that are useful toward helping teachers learn and grow.

WHAT KINDS OF INTERACTIONS WILL ENCOURAGE THE GROWTH OF MATHEMATICS TEACHERS?

In the absence of shared work, teachers' interactions are not collegial but friendly. Experience swapping, storytelling, sharing, assisting, and aiding are not examples of collegiality. In fact, there is a significant difference between congeniality and collegiality. Congeniality is about close-knit personal relationships, promoting a positive working environment. In a congenial environment, teachers may be professionally isolated in regard to their work. Collegiality, however, refers to the existence of high levels of collaboration among teachers, the product of teachers working together on a common project toward some common goal. According to Hamstra (1996, p. 32):

> This concept of joint work requires teachers to participate with one another. They do this because they recognize each other's contributions in order to succeed in their own work. Under this model, encounters among teachers are driven by the shared responsibility for the work of teaching.

Interactions that are collegial are those in which teachers engage voluntarily with the intention of improving their practice. Collegial interaction also builds a common core of experiences and knowledge, which serves to increase the effectiveness of communication as well as build professional relationships.

With collegiality comes mutual respect, shared values, cooperation, and specific conversations about teaching and learning. "Thus, while the product of exchange in isolated settings is often sympathy, the product of exchange in a collegial setting is often ideas" (Rosenholtz 1989, p. 378). The ideas, then, flow throughout the collegial setting, including in the lunchroom and parking lots. All this serves to encourage more teaching and learning and begets more collegiality.

The specific kinds of collegial interactions that lead to professional development and teacher change were defined by Little (1982, p. 331):

> Continuous professional development appears to be most surely and thoroughly achieved when:
>
> 1. Teachers engage in frequent, continuous, and increasingly concrete and precise talk about teaching practice
> 2. Teachers are frequently observed and provided with useful (if potentially frightening) critiques of their teaching
> 3. Teachers plan, design, research, evaluate, and prepare teaching materials together
> 4. Teachers teach each other the practice of teaching.

The first type of collegial interaction mentioned by Little is teacher talk of a very specific nature. It does not include casual conversations of a personal nature, nor does it include conversations involving generalities about teaching, mathematics classes, or even mathematics. It includes discussions about particular mathematical situations, pedagogical situations, or both. More specifically, it includes the kind of reflective discussions that help teachers evaluate experiences and share knowledge embedded in particular content (Sykes 1999). For example, a conversation might begin by asking whether students are able to transfer what they are learning about algebra tiles into symbolic form and then use strategies to build "bridges" between the manipulatives and the symbols. The job-embedded nature of these discussions reminds us that teachers, like students, learn best when they are solving problems that they have identified and that are in the context of their daily lives.

The second type of collegial interaction, being observed and critiqued, is seldom done but often can be fruitful. Teachers are most frequently observed by administrators or others in positions of authority, but it is nonthreatening observation by their peers that usually leads to more significant learning on the part of teachers (Jones et al. 1994). Getting meaningful feedback from a colleague who has observed a particular teaching episode gives teachers the opportunity to focus their analysis and reflection on particular needs of which they may be unaware. It also offers them a new source of insight from another perspective. For the teacher doing the observing, there is a unique opportunity to build an image of mathematics teaching that may not match his or her own.

The third form of collegial interaction cited as beneficial for both the professional status of teaching and the professionalism of the teacher is collaborative teacher research (Stevenson 1998). Whether it is a formal research study or preparing and teaching lessons, the idea is that mathematics teachers will find ways to work together to improve their teaching in an inquiry-based, problem-solving model. "Teacher engagement in research helps create a clientele for profession-wide knowledge while it also builds teachers' personal knowledge of students and learning in ways that are often transformatory for teaching" (National Commission on Teaching and America's Future 1996, p. 12). Hence, in the effort of improving their own teaching, mathematics teachers can contribute to the establishment and refinement of standard practices of the profession of teaching.

The fourth form of professional collegial interaction happens when teachers teach other teachers. Certainly one form of this can be seen at professional conferences. Every year, hundreds of sessions at conferences are led by mathematics teachers who share ideas with others. In a collegial school environment, similar episodes can occur in planned sessions, such as departmental meetings, as well as spontaneously through the everyday sharing of ideas.

> In the most adaptable schools, most staff, at one time or another, on some topic or task, are permitted and encouraged to play the role of instructor for others. In this way, the school makes maximum use of its own resources. (Little 1982, p. 331).

A common characteristic of the four kinds of collegial interactions discussed here is that they have the potential for establishing a common language "adequate to the complexity of teaching, capable of distinguishing one practice and its virtues from another, and capable of integrating large bodies of practice into distinct and sensible perspectives on the business of teaching" (Little 1982, p. 331).

Those aspects of collegial interaction that make it a useful and important tool for professional development are its concreteness, its precision, and the coherence of the shared language evolving from the interaction. This shared language, then, paves the way for a common understanding, shared values, shared goals, and fertile ground for learning by teachers (Hill 1995). The end result of this collegial interaction is teachers' growth and changes in how and what is being taught.

HOW CAN COLLEGIAL INTERACTION AMONG MATHEMATICS TEACHERS BE SUPPORTED?

Four "Pillars of Support for Collegial Interaction" emerged from a study of high school mathematics departments engaged in reform efforts (Taylor 2001). This framework is intended to be a tool for analyzing the capacity for supporting reform through collegial interaction, whether analyzing the current status of an organization or designing a future reform effort. The four pillars of support represent foundations for building the capacity to support collegial interaction: contextual catalysts, the identification of colleagues, the alignment of philosophies, and the ownership of interaction.

Contextual catalysts are those elements of the department, school, district, and beyond that affect the interactions among mathematics teachers in a school. The structures of the school day and school year, for instance, play a large role in determining whether mathematics teachers have opportunities for ongoing interactions. The physical structure of the teachers' workplace is also an important factor. Teachers who are separated by distance, located on different floors, and who have no common working space or no common working time are less likely to interact in significant ways. The leadership style in a school also affects collegial interactions. A democratic leader will create opportunities for the mathematics teachers to solve problems collectively, whereas a more bureaucratic leader will tend to tell teachers what work to accomplish. Other examples of contextual catalysts include turnover in the faculty, changes in state assessments, and larger reform efforts such as those envisioned in the NCTM *Standards*.

Identification of colleagues refers to the process by which teachers define themselves and select peers with whom they should work. For example, a high school teacher who identifies himself as a lower-level teacher, teaching prealgebra and applied mathematics courses, may not see much benefit in working with the "calculus teacher." Likewise, those who see themselves as aligned with the NCTM *Standards* may not view work done with a fellow mathematics teacher who values a back-to-basics approach as collegial interaction. The process of identifying colleagues, then, narrows the field of colleagues available for interaction that the teacher finds meaningful and significant.

Alignment of philosophies addresses the issue of whether the interaction in question is well aligned with the beliefs of the teachers involved. Asking a teacher who believes that calculators replace mathematical thinking to work on a project that implements technology in the classroom is not likely to lead to meaningful collegial interaction. A more subtle example would be that of a teacher who is asked to spend several days each month planning how the school will meet the district's requirement for "writing across the curriculum." Although he was not opposed to the concept, this particular teacher viewed these planning meetings as taking up time that should be dedicated to mathematics. If the work is not in line with the teacher's goals and beliefs, the work itself is viewed as wasteful; therefore, the interaction with colleagues is not viewed as meaningful.

Ownership of interaction, like the alignment of philosophies, refers to the teachers' view of the value of the interaction. Even if the interaction is with those a teacher has identified as colleagues and aligns with the teacher's beliefs, the interaction can still be seen as superfluous if the focus of the interaction is low on the teacher's priorities. For instance, teachers having trouble keeping up with planning may view just about any effort as wasteful. As a result, they are much less likely to engage in an activity with conviction and find significant growth from the interaction. However, if the teachers believe that they have chosen to engage in the effort, that it aligns with their beliefs, and that it is done with those whom they consider colleagues on the issue, they are likely to view that effort as meaningful collegial interaction.

The notion of "pillars of support" is intended to reflect the precarious nature of collegial interaction. If four pillars support a structure and one is removed, the entire structure may collapse. Similarly, if one of the four issues is not properly addressed, the collegial interaction, as well as the growth resulting from it, is unlikely to be significant for the mathematics teachers involved.

Two High Schools' Efforts at Reform through Collegial Interaction

Research into reform efforts gives more insight into the way the four pillars work. In the following section, two high schools, Dayville and Tartan High Schools (pseudonyms), are used as examples (Taylor 2001). On the surface, the efforts toward reform at the two schools looked very similar. Both of these large suburban high schools were engaged in reform efforts that required a substantial amount of collaboration among the members of their mathematics departments. The Dayville mathematics department was working on a grant that focused on improving its scores on a recently revised state assessment. This work included monthly meetings to share methods of intervention and to discuss the results. In particular, mathematics problems and assessment plans were discussed, and it was expected that teachers would try out these interventions before the next meeting. The results of each intervention task were discussed, trends in students' work and scores were identified, and then the tasks were refined.

The Tartan mathematics department was implementing "writing across the curriculum," a rewriting of curriculum objectives for all courses, and a partial implementation of a *Standards*-based curriculum. These activities took up all the professional development time available in the regular work schedule. The writing-across-the-curriculum project required teachers to develop specific math-related writing assignments to be embedded in each course. The rewriting of course objectives was in response to a grant that was done at the school level and required that each course syllabus be pared down to ten objectives. This was a schoolwide project that was handed to the mathematics department to accomplish. The partial implementation of a *Standards*-based curriculum was done by the prealgebra teachers, but it was a very low priority for the department and the school.

An examination of the "four pillars" involved in supporting collegial interaction shows significant differences in how collegial interaction was supported or deterred in these two mathematics departments (see fig. 18.1).

Contextual Catalysts

Contextual catalysts varied between the two schools. The administration at Dayville was very democratic. The teachers had a significant role in how the school was run, and the principal openly promoted the idea of teachers as researchers of their own work. In response to this philosophy, to changes in the state assessment, and to the NCTM call for reform, the Dayville mathematics teachers had made a collective decision to write a grant to work on improving their state assessment scores. They decided how their collaborative work would be structured and measured. The mathematics teachers at Tartan High School, however, believed that nearly all their collaboration efforts were spent on projects that were determined by the district or by the state. They reported that they would like to be working on different goals but did not have the time or latitude to do so.

Another important difference in the contextual catalysts was the department chairs. The Dayville mathematics chair had designed a very collegial environment in both the physical structure of the department office, which could seat all members around one long table, and the activities of the department. In contrast, the Tartan mathematics department chair had a self-proclaimed laissez-faire leadership style and the mathematics department office could not seat even two-thirds of the department at once.

Identification of Colleagues

Both the Dayville and the Tartan mathematics departments had multiple pockets of collegiality, as defined by peers they would choose to work with on teaching-related tasks. Collegial pockets in both departments were determined by course, beliefs, and proximity. The difference between the two efforts was that the Dayville teachers had more say about with whom they would collaborate. Tartan teachers were often in situations where they would work with people whom they identified as coworkers but not as colleagues. As a result, Dayville teachers were more likely to view the interac-

	Dayville High School (DHS)	Tartan High School (THS)
Contextual Catalysts	Department chair sought collegiality. Democratic administration	Department chair had laissez-faire leadership style. Bureaucratic administration
Identification of Colleagues	Teachers' collegial groups defined by course, by beliefs, and by proximity. Teachers could often choose with whom they worked.	Teachers' collegial groups defined by course, by beliefs, and by proximity. Teachers were less likely to be able to pick those with whom they were to interact.
Alignment of Philosophies	Since they chose what work to work on and how to work on it, the work was usually aligned with their philosophies.	Since they rarely chose what to work on or how to work on it, the work was not well aligned with their philosophies.
Ownership of Interaction	To a large extent, the teachers defined the interactions. Hence, they felt ownership.	Very little of their common work was defined by the teachers. As a result, it was seen as "extra" and not their work.
RESULTS	DHS mathematics teachers' interactions were genuinely collegial and led to professional growth and eventually to improved student performance.	THS mathematics teachers' collaboration was mostly coerced. This led to resentment and bitterness rather than professional growth.

Fig. 18.1. An overview of the "four pillars" in two high school mathematics departments

tion as meaningful, even when choosing beyond their established pockets of collegiality.

Alignment of Philosophies

Having more control over the focus of their efforts, the Dayville mathematics teachers tended to work on items aligned with their beliefs. The Tartan mathematics teachers had their goals and often the means of their collaborative work defined by others.

Ownership of Interaction

The Dayville mathematics teachers had collectively decided to engage in reform activities to meet their personal and departmental goals. As a result, they believed that these were *their* projects. The Tartan mathematics teachers felt that projects were imposed on them. They did not think that they had any time left to engage in collegial interactions that they would choose to initiate given the opportunity. Even when they would choose to work on a particular project, the nature of the work, such as meetings after school or meeting during teaching time, may not have been valued.

The results of these conditions were that the Dayville mathematics teachers were in a situation in which they were more likely to learn and grow. They engaged in meaningful collegial interaction opportunities that were comparable to the kind of inquiry-based learning they sought for their students. Most of the collaboration among the Tartan teachers, however, was not collegial in the truest sense. They were often working with peers beyond their self-identified collegial pockets. In those situations, they placed less value on the input of the peers with whom they were interacting. In addition, they often did not value the goals or the format of the collaboration. As a result, they were less likely to think of these interactions as significant learning experiences. The outcomes of these interactions, as measured by the students' opportunities to learn, varied accordingly. The Dayville mathematics teachers were implementing changes in their teaching that were based on their collegial interactions; the Tartan mathematics teachers were adding tasks they did not necessarily value to an already dense curriculum.

LESSONS LEARNED ABOUT SUPPORTING REFORM THROUGH COLLEGIAL INTERACTION

What should readers learn from this article and apply to their own situation? First, note that collegial interaction can be a useful tool for professional development. Second, recognize what kinds of interactions lead to professional growth and which ones do not. Third, understand that there are conditions that favor collegial interactions that lead to positive professional growth. The model for support set forth here is a starting point in assessing an organization's or project's capacity for promoting collegial interaction. By understanding the four pillars of support and applying them to their context, administrators, department chairs, and mathematics teachers can develop more effective professional development by creating an environment that encourages both individual and collective growth.

REFERENCES

Friel, Susan, and George Bright, eds. *Reflecting on Our Work: NSF Teacher Enhancement in K–6 Mathematics.* Lanham, Md.: University Press of America, 1997.

Hamstra, Janet. "A Teacher Network: Collegiality and Professionalism in a Learning Community." Ph.D. diss., University of California, Los Angeles, 1996. *Dissertation Abstracts International* 57, no. 06A (1996): 2342.

Hill, Donald. "The Strong Department: Building the Department as Learning Community." In *The Subjects in Question: Departmental Organization and the High School,* edited by Leslie Santee Siskin and Judith Warren Little, pp. 123–40. New York: Teachers College Press, 1995.

Jones, Graham A., Cheryl A. Lubinski, Jane O. Swafford, and Carol A. Thornton. "A Framework for the Professional Development of K–12 Mathematics Teachers." In *Professional Development for Teachers of Mathematics,* 1994 Yearbook of the National Council of Teachers of Mathematics (NCTM), edited by Douglas B. Aichele, pp. 23–36. Reston, Va.: NCTM, 1994.

Lappan, Glenda. "Revitalizing and Refocusing Our Efforts." *Journal for Research in Mathematics Education* 30 (November 1999): 568–78.

Little, Judith W. "Norms of Collegiality and Experimentation: Workplace Conditions of School Success." *American Educational Research Journal* 19 (1982): 325–40.

National Commission on Teaching and America's Future. *What Matters Most: Teaching for America's Future.* New York: National Commission on Teaching and America's Future, 1996.

National Council of Teachers of Mathematics (NCTM). *Curriculum and Evaluation Standards for School Mathematics.* Reston, Va.: NCTM, 1989.

———. *Professional Standards for Teaching Mathematics.* Reston, Va.: NCTM, 1991.

———. *Assessment Standards for School Mathematics.* Reston, Va.: NCTM, 1995.

———. *Principles and Standards for School Mathematics.* Reston, Va.: NCTM, 2000.

Rosenholtz, Susan J. *Teachers' Workplace: The Social Organization of the Schools.* White Plains, N.Y.: Longman, Inc, 1989.

Stevenson, Robert B. "Educational Practitioners' Use of Research: Expanding Conventional Understanding." In *Transforming Schools and Schools of Education,* edited by Stephen L. Jacobson, Catherine Emihovich, Jack Helfrich, Hugh G. Petrie, and Robert B. Stevenson, pp. 99–122. Thousand Oaks, Calif.: Corwin Press, 1998.

Stigler, James W., and James Hiebert. *The Teaching Gap.* New York: The Free Press, 1999.

Sykes, Gary. "Teacher and Student Learning: Strengthening Their Connection." In *Teaching as the Learning Profession,* edited by Linda Darling-Hammond and Gary Sykes, pp. 151–80. San Francisco: Jossey-Bass Publishers, 1999.

Taylor, P. Mark. "Collegial Interactions among Missouri High School Mathematics Teachers: Examining the Context of Reform." Ph.D. diss., University of Missouri—Columbia, 2001. *Dissertation Abstracts International* 64, no. 05A (2001): 1764.

19

Developing a Support System for Teacher Change in Mathematics Education

The Principal's Role

Patricia A. Jaberg
Cheryl A. Lubinski
Sigrid Aeschleman

THE principal plays a critical role in facilitating a shift in the environment of the mathematics classroom as teachers strive to develop and implement a mathematics curriculum based on children's thinking and reflecting the vision of the National Council of Teachers of Mathematics (NCTM) *Standards.* One principal of a rural Illinois school not only supported a project that precipitated change but also provided opportunities for the staff to communicate with colleagues, as well as with students, parents, and other teacher-educators.

Teachers are more likely to modify their teaching methods if they have access to knowledge about students' thinking (Peterson, Fennema, and Carpenter 1991), are offered time to reflect on and discuss their teaching style (Borko et al. 1997), and have support from colleagues (Lampert 1988). As early as 1991, NCTM recommended that administrators should also take an active role in supporting teachers of mathematics by accepting responsibility for

- involving teachers in designing professional development;
- establishing parental communication and involvement in relation to quality mathematics programs;

The preparation of this article was supported in part by a grant from the Dwight D. Eisenhower Mathematics and Science Education Act for "Instructional Decision Making and Mathematics: Using Students' Thinking." Any opinions expressed herein are those of the authors and do not necessarily reflect the views of the Department of Education.

- providing a support system for novice and experienced teachers to encourage profession growth;
- supporting collaboration among teachers in order to improve the teaching of mathematics;
- providing sufficient resources to support the teaching and learning of mathematics;
- establishing adequate reward systems to facilitate excellence in teaching mathematics.

THE PROJECT

Our staff development project was based on the assumptions that (1) teachers' instructional decisions would be better if teachers understood their students' thinking, and (2) through support and collaboration, teachers could develop effective learning environments. This meant that any ideas introduced during the project had to consider the development of students' thinking. Furthermore, an implicit goal of the project was to provide time for teachers to communicate about their teaching and their students' learning of mathematics and to develop a collegial support network around instructional practices. This project was directed by a university-level mathematics educator and included the principal and the entire school staff of eleven teachers in grades K–4 and three classroom aides, all female.

All participants attended six spring sessions, an intensive one-week summer in-service session, and a culminating fall session. Additionally, the university educator visited each participant's classroom during both the spring and the fall to conduct an assessment of a lesson. The spring assessment was done to collect information on current instructional practices. In the fall, the assessment was done to collect information on instructional practices after an intervention. The fall lesson was compared to the spring lesson in order to determine the degree to which teachers had modified their instructional decisions with respect to the use of students' thinking. Instructional practices were assessed in three ways. Both teachers and aides were asked (1) to respond to questions involving the purpose, content, and instructional decisions considered prior to teaching a mathematics lesson; (2) to videotape the lesson; and then (3) to select three lesson segments from the video and discuss the decisions they made during each of the segments in relation to their students' responses.

Content analysis of the lesson plans and the videotaped mathematics lessons suggested that teachers did alter their instructional practices and decision making. The lessons were evaluated according to (1) how much time the teacher talked during the lesson relative to the amount of time the students talked during the lesson and (2) what the teacher said during the lesson. Results (see fig. 19.1) showed that teachers spent significantly less time talking and explaining following the intervention than they did prior to the intervention.

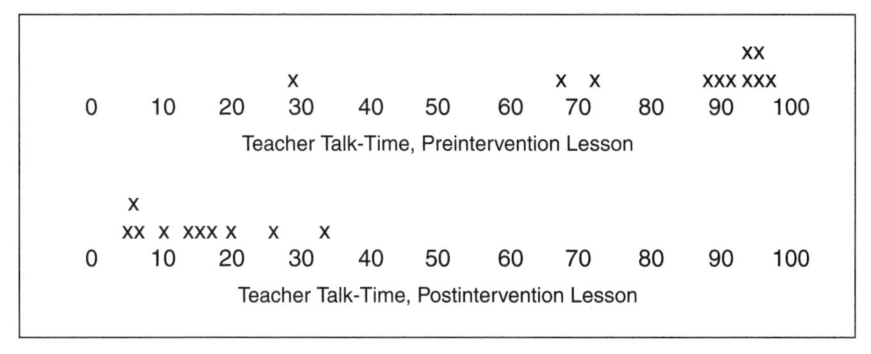

Fig. 19.1. Percent of class time defined as teacher-talk time (adapted and reprinted from Lubinski and Jaberg 1997)

Additionally, the results indicated that what the teachers said also changed. Preintervention lessons were characterized by questions on content and knowledge that afforded little opportunity for interaction and discussion. The curriculum was textbook-driven, and lessons were restricted to topics and numbers that the textbook suggested. After the intervention, the teacher's manual received decreased attention, and the lessons were characterized by higher-level questioning that encouraged students to explain their thinking. Teachers reported that they spent more time talking about the mathematics and asking for multiple solutions to problems. Teachers reported that they were more aware of individual strategies and that hearing students discuss mathematics was exciting and enlightening. Two examples illustrate these changes.

In preparation for performing subtraction with double digits, a second-grade teacher's preintervention lesson focused on an activity that practiced regrouping. After she demonstrated that 18 ones is equal to 1 ten and 8 ones, she asked, "What will we do for subtraction? We'll take away or trade back a ten for ones." Pairs of children were directed to demonstrate this task with straws. The focus was on learning and practicing the procedure. In contrast, during the postintervention lesson, children were encouraged to reflect on their own thinking. Multiple solutions to problems were encouraged and discussed. During the lesson, the teacher exclaimed, "We found three ways to do this problem!" The postintervention lesson involved addition with three-digit numbers, such as 134 and 246. The preintervention regrouping lesson was restricted to numbers less than 20. After the intervention, the teacher was obviously not restricted by the content of the textbook, and her instructional decisions reflected a belief that second-grade students could deal with larger numbers with understanding.

Influenced by the textbook, a third-grade teacher used beans and cups to illustrate the numbers 2 and 3 as divisors. She reported that decisions related to the preintervention lesson were influenced by "security—go strictly by the book, nice and safe." In contrast, this same teacher began her postintervention lesson by posing this problem: "Think of the number 100. Go all the way down to zero by subtracting the same number. First, think of one number you can subtract four times to get to zero." She later asked her students to find all possible combinations of numbers. Afterward, as she reflected on her lesson, she wrote in her notes that the children didn't respond as she had anticipated they would:

> I thought the students would do subtraction, but they didn't. *All* the students chose to use repeated addition, which was fine because we began discussing multiplication. We learned that repeated addition is the same as multiplying. The students were able to tell me that if we switched the numbers around, the problem would still be the same. For example, $4 \times 5 = 5 \times 4$. The students were able to relate their knowledge of addition to multiplication. The problem helped the students become enthusiastic about multiplication. It is going to be exciting to explore multiplication. [We became] so involved in the problem that we ran out of time.

This example illustrates how instruction was built on students' previous knowledge to develop mathematical understanding. The teacher asked the students to discuss what they knew about addition and multiplication, as well as how it was related to the problem posed.

Data collected during the fall semester shows that students were learning to represent their thinking symbolically through problem solving and classroom discourse. For example, in her preintervention lesson, a first-grade teacher emphasized memorization as a technique to learn subtraction facts related to 12. In contrast, her postintervention lesson consisted of three word problems, chosen "to help make the connection with computation." She exhibited an awareness of her students' levels of thinking and used a variety of problem types with challenging numbers. She asked the students to write some type of representation for their solution. These symbolic representations reflected their strategies and thinking, and when a student was unable to do this, the teacher questioned and discussed the solution with the student in order to help clarify his or her thinking to the point at which representation emerged.

Thus, several degrees of positive results were obtained from both staff and students in relation to beliefs, practices, and learner outcomes. Communication was more effective between students and teachers than it had been prior to the intervention, and it was used to assess learning and to plan future instruction.

Making a Change

It is important to note that change often involves struggle; however, the teachers in this project did not struggle to the extent we have seen others struggle. First, it was the teaching staff who recognized a need to change and asked for staff development in the area of mathematics. Second, the teaching staff had ongoing support both from the university mathematics educator and their principal. We believe these two factors contributed to our perception that struggle was minimized during the transition from traditional to *Standards*-based instruction. Also, we have noted that the degree of change varied among teachers.

The Role of the Principal

School administrators play an important part in any program seeking to change teachers' practice. Perhaps the most important of these administrators is the school principal. It is the school principal who allocates resources, conducts evaluations, and interacts with the teachers most often. It is also the school principal who is most responsible for establishing a climate that is supportive of any effort at change (NCTM 1991, 2000).

Mrs. A was the principal at a rural elementary school in central Illinois. She and the entire staff participated in the project. By participating herself, Mrs. A outwardly communicated to the staff that change in teaching mathematics was necessary and that she would support them in their efforts by keeping herself informed of their work. She perceived her supportive role, based on effective communication, as instrumental from the inception of the project to well beyond the formal ending of the project.

Support for School and University Communication

The impetus for change originated with one of Mrs. A's teachers, Mrs. S, who had attended a summer workshop at a local university on teachers' decision making and students' thinking in teaching mathematics. Mrs. S communicated her learning experience to Mrs. A, and they discussed the possibility of offering the staff the same information to which she had had access during the workshop. Shortly thereafter, Mrs. A conducted a faculty meeting at which Mrs. S explained a proposed university-partnership project to the teachers and aides. It was important to Mrs. A that Mrs. S communicate the information, which included a consideration of the effects on their personal time, their willingness to make changes, the effects on the curriculum, and the ramifications of part of the staff being involved versus the total group. (Mrs. A was concerned that if only part of the staff was involved, instructional decisions and curricular planning would be extremely difficult.

Further, follow-through involving the children's progress within two instructional approaches, one using students' thinking and the other planning only with the textbook, would be difficult. Thus, the entire staff decided to participate.)

In a letter to the university teacher-educator verifying that the staff would participate in the project, Mrs. A indicated that the school district would commit to purchasing the *Standards* for each teacher and up to $300 dollars of mathematics manipulatives for the students' use. After the staff completed the summer session, Mrs. A asked each teacher to submit a list of materials she would like for her classroom.

Communicating with Parents and the Community to Build Support

After the staff had decided to participate, Mrs. A sought the support and approval of the school board. At a regular meeting of the board of education, she described the project and supplied a rationale for change. Board members were informed that change was anticipated in both the mathematics curriculum and instructional practices. For example, change might occur in the instructional presentation, sequence, and assessment procedures. They formally gave their support to the project. Information was communicated to both parents and students about the upcoming project through school board minutes and both the district and the school newsletters. As the project progressed, Mrs. A occasionally publicized the mathematics activities of the school in both the local newspaper and the building reports, highlighting the positive effects of the project.

Before school began in the fall, Mrs. A conducted a parent orientation night at which she communicated information about the progress of the project to parents. Teachers explained their classroom procedures, practices, and expectations. Communication between parents and staff was further enhanced when an opportunity for questions was presented and handouts were given to parents. Following the presentation, a number of interested parents remained to ask additional questions, and the principal was able to further clarify the project's objectives, the rationale for change, and the implications for students. Thus, communication between parents and school was encouraged, and support for changes in mathematical instruction was generated.

SUPPORT FOR PROFESSIONAL COMMUNICATION

All participants in the project were given individual copies of the *Curriculum and Evaluation Standards for School Mathematics* (NCTM 1989). As a participant, Mrs. A familiarized herself with the content of this document.

Throughout the project, she believed it was important to expose the staff to research-based articles to reinforce concepts covered in the seminars. She disseminated such articles to the staff on a regular basis. She also frequently offered opportunities for the staff to discuss mathematics during grade-level meetings, individual consultations, and faculty meetings. Communication focused on the project and its impact on curriculum, materials, pacing, and student achievement.

Mrs. A had to change her own thinking about the teaching and learning of mathematics because previously she had required evidence of student learning through assessment instruments such as criterion-referenced tests and checklists that corresponded to the chapters in the mathematics textbook. Now, she accepted the flexibility that had to occur in the pacing of the curriculum because of the changes being made in the sequencing of mathematical concepts. For example, a mid-year textbook test typically administered in January might not be given until April or early May when the teachers perceived that their students were ready to be assessed on particular concepts.

Supporting Change

Mrs. A found herself continually having to reinforce the changes that the staff was making because they were asking for suggestions and wanting to know how she thought things were going in their classrooms. In the beginning, they appeared frustrated until such issues as pacing and sequencing of topics were settled. Also, they were at first concerned about how to handle questions from parents. For example, "Why has the number of mathematics worksheets coming home decreased?" Teachers were unsure about how they would document students' achievement on the report card, since grades still had to be assigned. Mrs. A believed she offered a support system for instructional change by

- verbally supporting the project;
- presenting regular updates to the board of education and the community, highlighting the staff's successes and time involvement;
- using opportunities to relate the success of the project to the general focus and mission of the school and district;
- providing a flexible and "safe" environment for the teachers to try new ideas and change established practices;
- meeting with the novice teachers on a weekly basis to give an opportunity to discuss issues, problems, successes, and student concerns;
- using the project as an example of effective teaching strategies and opportunities for professional growth;
- remaining flexible with the structure of lesson plans, learning-outcome data, and assessments submitted by the teachers.

Additionally, Mrs. A believed she provided support for teachers' self-evaluation and growth in analyzing, evaluating, and improving their teaching by

- encouraging teachers to participate in peer-coaching activities by observing and discussing one another's lessons;
- giving feedback to teachers on a regular basis in both verbal and written form, particularly noting positive changes that were occurring;
- making teachers, parents, and the board of education aware of successes in the classroom (e.g., constructing a "Principal's Proud Board" in the hallway);
- videotaping preintervention and postintervention mathematics lessons;
- highlighting the work of teachers and their students (e.g., showing videotaped lessons at administrators' workshops);
- writing reactions to each of the seminars and the workshop in the form of an ongoing journal in order to determine needed changes in the project model.

Furthermore, Mrs. A allowed time for teachers to discuss their mathematics curriculum during monthly faculty meetings, grade-level meetings, and individual consultations. She allocated time to observe teachers formally for purposes of evaluation during their mathematics periods.

Ongoing Support

Both during the project and well beyond the end of the project, Mrs. A continued to be sensitive to her staff's needs "as they perceive them" for materials and time, as well as the concerns of the students. She held high expectations for students and staff and highlighted mathematical successes. She continued to promote a school climate that was professional, caring, and responsible; she expected such characteristics from the staff as well. Mrs. A encouraged opportunities for professional collaboration. She substitute-taught for her own teachers occasionally, while they observed and coached one another; she gave "free period coupons," which teachers used to have the principal teach one period of their class. Mrs. A fostered leadership qualities among the staff themselves by using such phrases as "You make the decision," "What do you think?" "Meet with So-and-so, and if you need me or want me there, let me know." Finally, she continued to write supportive, encouraging notes to the teachers.

STRATEGIES FOR SUPPORTING TEACHER CHANGE

This article discussed the important role the principal played in a project involving teacher change and mathematics. It is important to note that change was just beginning to occur at this school, and the principal was aware of this. She continued to offer opportunities for teachers to communicate about how they were developing their knowledge of mathematics and their students' thinking about mathematics.

Following this project, six teachers chose to participate in a five-year National Science Foundation grant designed to expand their understanding of students' thinking and build their repertoire of teaching strategies. Mrs. A believed that without her support in the initial project, these teachers would not have been as motivated to participate in an extension project. When asked to list things she did to help make the project a success, she responded that it was necessary to communicate to the staff that she

1. had confidence in them and was supportive of their learning efforts;
2. would continue to generate support from parents, the school board, and the community;
3. would allocate funds for resources and materials that the teachers believe necessary in order to teach effectively;
4. would continue to focus on student learning outcomes;
5. would offer her own time to assist teachers in activities such as taping lessons for assessment and discussion purposes;
6. would be flexible in regard to curriculum goals and lesson plans;
7. would arrange time for staff to discuss the progress of their mathematics instruction.

Discussion

Mrs. A's role as principal reflected many of NCTM's (2000) recommendations. By arranging time for the teachers to reflect and collaborate within the school day, she created a work environment that was "conducive to productive professional interactions" (NCTM 2000, p. 376). Mrs. A offered opportunities that strengthened the focus on mathematics and showed support for "the improvement of mathematics education by establishing effective processes for the analysis and selection of instructional materials" (p. 377). By becoming part of the project herself, Mrs. A gained a greater understanding of the goals of mathematics instruction and understood the challenges of creating an environment that promoted learning mathematics with understanding.

Administrators can positively affect the quality of mathematics education by supporting the professional growth of mathematics teachers. This support can be accomplished by arranging for meaningful professional development workshops, creating schedules so that collaboration with colleagues is part of the school day, and shaping "work environments so that they are conducive to productive professional interactions" (NCTM 2000, p. 376). Administrators—and principals in particular—have powerful opportunities to strengthen the focus on mathematics and support the improvement of mathematics education to achieve the goal of mathematical power for all students.

References

Borko, Hilda, Vicky Mayfield, Scott Marion, Roberta Flexer, and Kate Cumbo. "Teachers' Developing Ideas and Practices about Mathematics Performance Assessment: Successes, Stumbling Blocks, and Implications for Professional Development." *Teaching & Teacher Education* 13 (1997): 259–78.

Lampert, Magdalene. "What Can Research on Teacher Education Tell Us about Improving Quality in Mathematics Education?" *Teaching & Teacher Education* 4 (1988): 157–70.

Lubinski, Cheryl A., and Patricia A. Jaberg. "Teacher Change and Mathematics K–4: Developing a Theoretical Perspective." In *Mathematics Teachers in Transition*, edited by Elizabeth Fennema and Barbara S. Nelson, pp. 223–54. Mahwah, N.J.: Lawrence Erlbaum Associates, 1997.

National Council of Teachers of Mathematics (NCTM). *Curriculum and Evaluation Standards for School Mathematics.* Reston, Va.: NCTM, 1989.

———. *Professional Standards for Teaching Mathematics.* Reston, Va.: NCTM, 1991.

———. *Principles and Standards for School Mathematics.* Reston, Va.: NCTM, 2000.

Peterson, Penelope L., Elizabeth Fennema, and Thomas P. Carpenter. "Teachers' Knowledge of Students' Mathematics Problem-Solving Knowledge." In *Advances in Research on Teaching*, Vol. 2, edited by Jere Brophy, pp. 49–86. Greenwich, Conn.: JAI Press, 1991.

20

A Tool for the Teaching Principle

Professional Development *through* Assessment

Sandra K. Wilcox

Elizabeth M. Jones

An important feature of the Teaching Principle in *Principles and Standards for School Mathematics* (National Council of Teachers of Mathematics [NCTM] 2000) is the recognition that teachers need opportunities and resources for ongoing, high-quality professional development. This stance is grounded in a belief that the curriculum and discourse of professional development must shift away from simply giving teachers things to do in their classroom toward providing support for more reflective approaches to the practices of teaching.

Why Tools?

People are much more effective at most activities if they have well-engineered tools to use.[1] Consider a shovel. Though simple in its design, it greatly improves a gardener's ability to move dirt, dig holes, and so on. A well-designed shovel is efficient and cost-effective. Most people can use it at some basic level, but a skilled user can execute the intended job at a new level. Like all good tools, the shovel has two ends, one for the user's hands and the other to perform the task. To generalize, an ideal tool

The work reported on here has been supported by the following grants from the National Science Foundation: ESI9612492 and ESI9726403. The opinions expressed in these materials do not necessarily represent the position, policy, or endorsement of the Foundation.

1. We are grateful to our colleagues Hugh Burkhardt, Alan Schoenfeld, Jim Ridgway, Phil Daro, and Mark St. John for this conceptualization of tools for change in mathematics education. The concept and the examples that follow appear in the proposal "Facilitating Mathematics Education Reform: Developing a Toolkit for Change Agents" (Mathematics Assessment Resource Service, 2001).

- addresses a real problem;
- is efficient, well designed, and cost-effective and has a specific function;
- extends the capabilities of both novice users and experts to get the job done;
- is robust, flexible, long lasting, and capable of assisting in a range of tasks.

Standards documents (e.g., NCTM 1989, 1995, 2000) are a prime example of a tool. The problem they seek to solve is how to know what to emphasize in mathematics education and at what level of accomplishment. The *Standards* exemplify the expectations for students' learning. They provide a framework for most teachers and districts to examine their curricula and for experts to develop state and district standards and grade-level benchmarks. One need only look to the development of state and local standards across the country since 1989, and to their influence on curricula and assessment, to judge the power of this tool to respond to a real problem.

Standards-based curricula are model tools. The problem they seek to solve is what to teach. The NSF-funded curricula provide exemplary, standards-based instructional materials and substantial support to the teachers implementing them. There is mounting evidence that in the hands of typical and expert teachers, these materials enhance the opportunities for students to engage with powerful mathematics. Teachers can do things they otherwise could not, and their students learn more (Briars 2001; Senk and Thompson 2003).

Standards-based, high-quality performance assessments are yet another tool. The problem they seek to solve is how to assess in a manner consistent with new goals and standards. Balanced assessments (e.g., Mathematics Assessment Resource Service 2001a, 2001b, 2002) target standards-based competencies in a way that traditional assessments do not. Performance assessments assist in a range of work. Tasks are used for professional development for typical teachers. Experts use them to reshape the forms and purposes of assessment practices in standards-based classrooms. As a result, teachers' knowledge about, and capacities to use, these assessments are enhanced, and students get to show what they know and what they can do (Ridgway et al. 2000).

Professional development is at the heart of systemic reform in mathematics education. Forms of professional development that focus on teachers' learning and continual advancement of their professional practice are the basis of another kind of tool. The problem this tool seeks to solve is what skills and understandings enable teachers to create standards-based classrooms. Below we describe such a tool. In this instance, professional development through assessment uses assessment as a powerful stimulus for broader development of capacity to teach for understanding.

Using Assessment as a Tool for Professional Development

Balanced Assessment in Mathematics is a workshop series, designed by the Mathematics Assessment Resource Service (MARS), for teacher-leaders and others who support the professional development of mathematics teachers in grades 3–10. The series is underpinned by a set of assumptions about assessment. High-quality assessment

- is an ongoing classroom activity aimed at gathering information in multiple ways about what students are coming to understand;
- supports the learning of mathematics that counts by focusing on important mathematical ideas, concepts, and processes;
- is balanced, covering the broad range of Content and Process Standards specified in *Principles and Standards for School Mathematics* (NCTM 2000);
- is fair to students, giving them opportunities to show what they know and can do;
- is of such high quality that students and teachers learn from them, so that assessment time also serves as instructional time;
- is practical so that assessment activities fit within the constraints of classroom daily life;
- provides useful information to teachers, so they can monitor and adjust their own teaching; to students and parents, so they can see where students are doing well and where they need more work.

All activities of the series have a high-quality performance assessment task as a central component. We elaborate on three themes that organize the learning activities of the series—understanding students' understanding, linking assessment with teaching and learning, and scoring what we value—with the T-Shirt Design task (Balanced Assessment Project 1999, pp. 106–7; see fig. 20.1) from the BA-MARS collection.

Understanding Students' Understanding

Activities built around the theme "understanding students' understanding" are designed to give teachers experiences to deepen and broaden their perspectives on assessment of students' mathematical skills, concepts, and processes as revealed in written work on a performance task. The activities aim to develop teachers' capacities to recognize the core elements of performance of a task, appreciate the multiple ways in which one can reason about and approach a task, and analyze students' responses to a task for evidence of understanding.

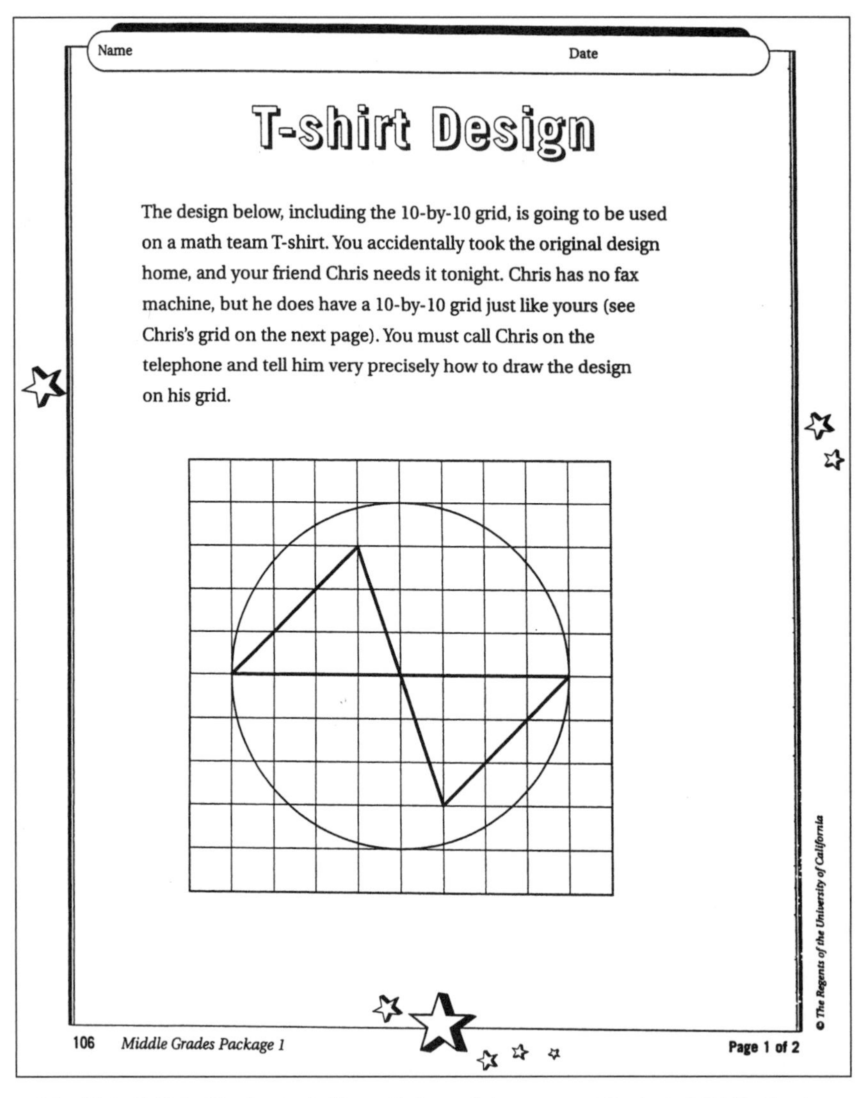

Name Date

T-shirt Design

The design below, including the 10-by-10 grid, is going to be used on a math team T-shirt. You accidentally took the original design home, and your friend Chris needs it tonight. Chris has no fax machine, but he does have a 10-by-10 grid just like yours (see Chris's grid on the next page). You must call Chris on the telephone and tell him very precisely how to draw the design on his grid.

106 *Middle Grades Package 1* Page 1 of 2

Fig. 20.1. T-Shirt Design task (from Balanced Assessment Project, *Middle Grades Assessment Package 1*, pp. 106–7), White Plains, N.Y.: Dale Seymour Publications, ©1999; used with permission). (*Figure continues in next page.*)

Working and reflecting on the task

The activity begins with teachers first working the task themselves. As they work, they consider the following questions:

- What are the "core elements of performance"?

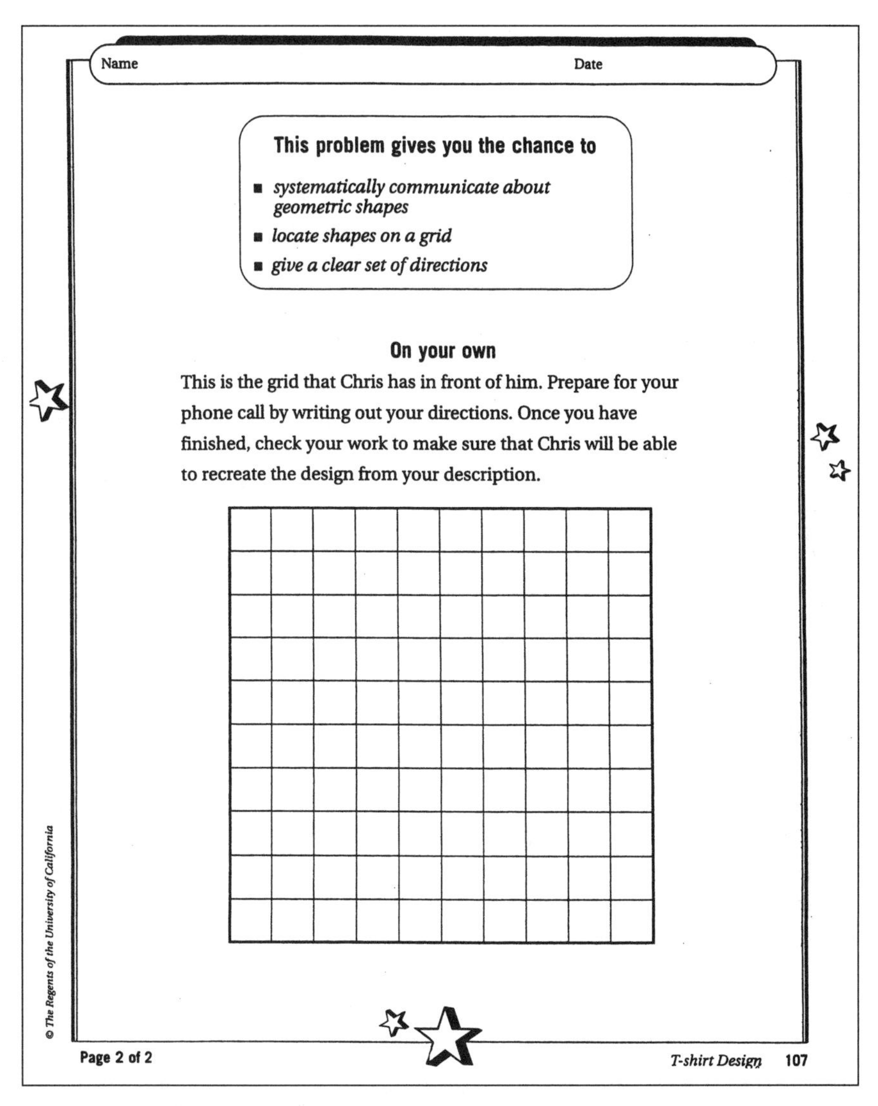
Name Date

This problem gives you the chance to

- *systematically communicate about geometric shapes*
- *locate shapes on a grid*
- *give a clear set of directions*

On your own

This is the grid that Chris has in front of him. Prepare for your phone call by writing out your directions. Once you have finished, check your work to make sure that Chris will be able to recreate the design from your description.

© The Regents of the University of California

Page 2 of 2 *T-shirt Design* 107

Fig. 20.1. T-Shirt Design task (*continued*)(from Balanced Assessment Project, *Middle Grades Assessment Package 1*, pp. 106–07), White Plains, N.Y.: Dale Seymour Publications, ©1999; used with permission).

- What could the task potentially reveal about students' understanding of content, problem solving and reasoning, and communication?

After teachers write their set of directions for T-Shirt Design, they read them aloud to others in their small group. This gives an opportunity to hear a variety of systems for locating the design on the grid and judge whether directions are complete and clear. Methods we have seen teachers use include the following:

- Cartesian coordinates, in either one or four quadrants
- Label one axis with letters and the other with numbers and give coordinates of key points in the form (*a*, 5)
- Directional—left, right, up, down
- Measurement (e.g., draw a circle with a radius of 4 centimeters)
- Rotation of one triangle to get the other
- A check (e.g., "When you are done, you should have a circle with a tilted bow tie in the center.")

Teachers discuss the "core elements of performance"—what a successful response must contain. In this task, students must give a clear set of directions, using an economic system for locating points that can accurately place all parts of the design on the grid. No particular system is privileged, and invented systems may be accepted. A successful solution does not require students to use particular vocabulary (e.g., *diameter, radius*), although the task has the potential to reveal whether students are disposed to use such terms.

Then teachers consider these questions:

- What do you think students are likely to do with this task?
- Why would a teacher want to use a task like this?
- Where might this task appropriately fit within your curriculum?

Examining students' written work on a task

Teachers closely examine samples of anonymous students' written work, which have been carefully chosen to engage teachers in lively and thoughtful conversations about students' understanding. The following questions help frame their analysis:

- What does the student seem to understand, and what is the evidence?
- What does the student seem to be struggling with? What does she seem to be confused about? Where does he show partial knowledge? What is the evidence?
- What logic might lead a student to respond incorrectly?

Often teachers have a tendency to look at students' work for right and wrong answers and to focus on what students have done incorrectly. Beginning with an analysis of what they think students do understand, and of what counts as evidence, shifts teachers' attention away from a singular focus on students' "deficits."

Figures 20.2 and 20.3 are two samples of students' work on the T-Shirt Design task. Teachers' analyses of these two responses tend to include the following.

The design below, including the 10-by-10 grid, is going to be used on a math team T-shirt. You accidentally took the original design home, and your friend Chris needs it tonight. Chris has no fax machine, but he does have a 10-by-10 grid just like yours (see Chris' grid on the next page). You must call Chris on the telephone and tell him very precisely how to draw the design on his grid.

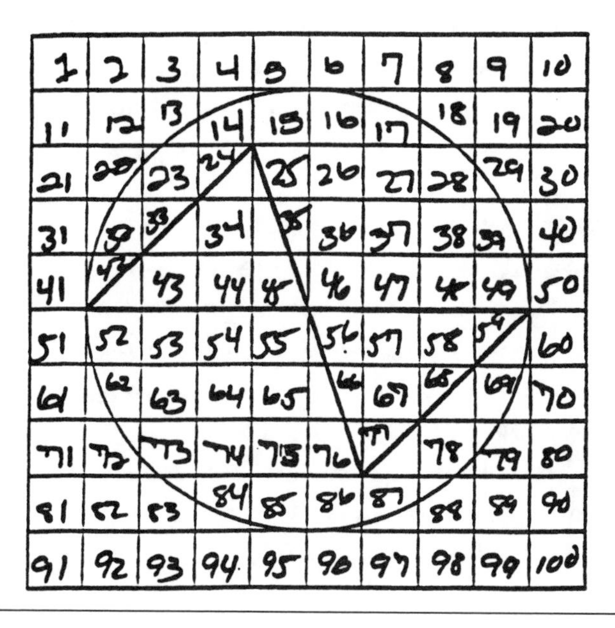

Fig. 20.2. Sasha's directions. (*Continued on next page*)

Sasha uses a rather unconventional system. He tells Chris to number every square on the grid, although he neglects to say precisely in what order. Then he gives directions by locating critical points at the intersection of four numbered squares on the grid. To locate one point on the zigzag, he identifies "… the center of the boxes 14, 15, 24, 25…." This system is used throughout, including setting boundaries on the placement of the circle. Sasha also combines this system with measurements. Although teachers find Sasha's system unusual, they note its orderliness and the place where he gives a check for Chris: "You should have a circle that is 4 inches wide and 4 inches tall. If you have a line from the top of the circle to the bottom of the circle, erase it." Teachers tend to agree that it is a system that is likely to result in the accurate placement of the design on the grid.

1. Number them 1 to 100 starting from
the top left-hand corner
2. At the line seperating 42 from 52
make a 4-inch line going from
boxes 42 and 52 across to 49 and 59
3. Make a Qaurter of a circle starting
at the left edge of the middle line going to
the center of 5, 6, 15, and 16.
4. make another Qaurter of a Circle
Starting at the center of 5, 6, 15, and 16
and ending at the center of 49, 50, 59, and
60.
5. make another Qaurter Of a Circle
Starting at the center of 49, 50, 59, and
60 and ending at the center of 85, 86, 95, and
6. Make another Qaurter of a Circle Start
at the Center of 85, 86, 95, and 96 and ending at
the Center of 41, 42, 51, and 52
which is where you started. You should
now have a Circle that is 4-inches
wide and 4-inches tall. If you have

Fig. 20.2. Sasha's directions (*continued*)

Danielle uses a quite different system. She locates the circle by giving its diameter (although she does not use the word) as the number of units in the middle of the grid. To locate the zigzag, she accurately describes lines at a 45-degree angle, and the correct placement of one of the slanted lines is likely. Her directions are ordered, and her system has the potential to locate all parts of the design. But it is less clear from her directions exactly how to draw the "straight line going left 45° going down from the right side of your straight line."

Taking up issues that emerge

Teachers are generally surprised to learn that large numbers of middle grades students do not apply conventional coordinates to this problem, even following a unit of instruction on coordinates and graphing. On further

have a line from the top of the circle
to the bottom of the circle
erase it.
7. make a line from the center of boxes
14, 15, 24, and 25 to the center of boxes
76, 77, 86, and 87.
8. Now make a line going from the
centers of the boxes 76, 77, 86 and 87.
to the right-hand end of the
line that you made in the beginning
9. Make a line connecting the
the line that begins at the
center of 14, 15, 24, 25 to the
left-hand end of the line
you made in the beginning.

Fig. 20.2. Sasha's directions (*continued*)

First draw a circle that is 8 units tall and 8 units wide in the middle of your grid. Then make a horizontal line that is 8 units long across the center of your circle. Next draw a straight line going right 45° going up three units from the left end of your straight line. after that draw a straight line going left 45° going down from the right side of your straight line. Then connect the ends of you're two 45° lines with a straight line.

Fig. 20.3. Danielle's directions

consideration, they speculate that what might account for this is that the T-Shirt Design task is quite different from the types of problems students typically encounter when working with coordinate graphing. This analysis leads to further discussion about whether and when students should be expected to use conventional Cartesian coordinates in this type of problem. It provides an example of the notion of students developing flexible knowledge—being able to recognize and bring to bear known mathematics to unfamiliar or nonroutine problem situations.

At the conclusion of this session, teachers are given an assignment to try the task in their own classrooms and bring three samples of students' work to the next workshop session. They are urged to select pieces that contain some puzzling elements that they have trouble interpreting, or that provide some new insights into a student's sense making.

Linking Assessment with Teaching and Learning

Assessment is a pivotal link between teaching and learning. The NCTM *Standards* documents (1989, 1991, 1995, 2000) emphasize the importance of assessment as ongoing activity in classrooms to help teachers monitor students' progress toward desired learning goals, judge the usefulness of learning activities they provide for their students, and assess and adapt their own instruction. This theme is taken up in several ways.

Using assessment to guide teaching

When teachers bring samples of their own students' work, they work with a partner. Each teacher describes a bit about the circumstances under which she or he gave the task—who the students are, whether and when the class had been working with the ideas embedded in the task, and what the teacher hoped to learn. After they examine together the samples of work, the teacher talks about what most students were able to do with the task and where they had some difficulties. On the basis of the analysis of where the class is strong and where they are weak, the partners consider what further learning activities the teacher might use to help students where their knowledge and skills are underdeveloped and how to select among the possibilities.

The question they consider is where they might go next in instruction. "Next" can mean what the teacher might do immediately as some follow-up activities. It can mean recognizing that some of the elements of performance in the task will be returned to in due course later in the year, and so the teacher makes note to come back to this task and issues it raised. Or "next" can mean how the teacher might teach some of the core mathematics to a new group of students the following year.

In one instance, a teacher had used the T-Shirt task following a unit on coordinate graphing. She was completely surprised that not a single student called on their knowledge of Cartesian coordinates to tackle the problem. In a follow-up discussion, she reflected on why students had not seen the connection between the ideas they had been working on and this new problem situation. She speculated that the context of the T-Shirt problem was quite different from any problems they had worked with in the unit. This led her to think more generally about providing tasks during instruction where students are called on to apply their knowledge to nonroutine, unfamiliar problem situations.

Orchestrating the work of students

Teachers must decide "how to support students without taking over the process of thinking for them and thus eliminating the challenge" (NCTM 2000, p. 19). For the T-Shirt Design task, we have another activity designed around a teacher's creative use of the task for assessment, peer feedback, and revision. Linda is a teacher of seventh graders in an urban middle school. Her orchestration of students' work is a powerful example of how "well-chosen tasks can pique students' curiosity and draw them into mathematics" (NCTM 2000, p. 18). She used the task and students' written directions in the following ways.

Day 1, first-hour class. Linda distributed the task and had a student read the directions. She gave them the time they needed to write their directions, and then she collected their work. Figure 20.4 shows the directions that Aaron wrote.

Day 1, second-hour class. Linda distributed a set of directions from the first-hour class to each student in the second-hour class. She also gave each student a blank grid. She told them they had directions for making a design on the grid. She instructed them to read the directions on their paper carefully, interpret them, and draw the design that they thought the directions

First of all make a circle inside
the grid by placing a dot on the
dead center line. Count 4 lines
up, 4 lines down and 4 lines from
side to side From the center dot, then
make dots on them (make sure your dots
are on the center line.)
Draw a line connecting the 2 side doTS
going From LeFt to Right. Then starting Back
up to The top dot, count (1 one) line
over and 1 line down.

Fig. 20.4. Aaron's directions

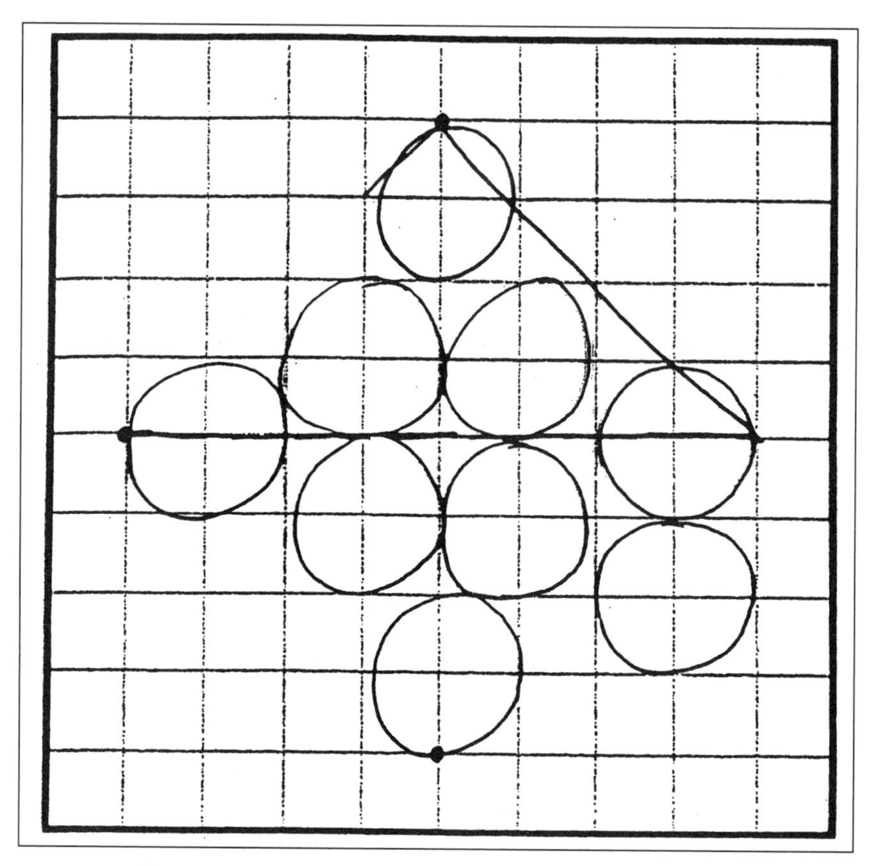

Fig. 20.5. Drawing made by a student using Aaron's directions

communicated. When they had made their best attempt at the drawing, she collected each set of directions with the drawing clipped to it. Figure 20.5 shows the drawing that a student made from Aaron's directions.

Day 2, First-Hour class. Linda returned each set of directions and the design made from it to the writer of the directions. She told them to look carefully at the design made by another student and compare it with their directions and the design in the task. Then she said they needed to revise their directions, taking into account the way another student may have interpreted the initial directions, with the aim of making the revised directions as clear and complete as possible. Figure 20.6 shows the revised set of directions that Aaron wrote after examining the drawing the second-hour student had made from his original directions.

Day 2, Third-Hour class. Linda gave the revised directions and a blank grid to students in yet another class who were unfamiliar with the design. She instructed them to read the directions, interpret them, and draw the

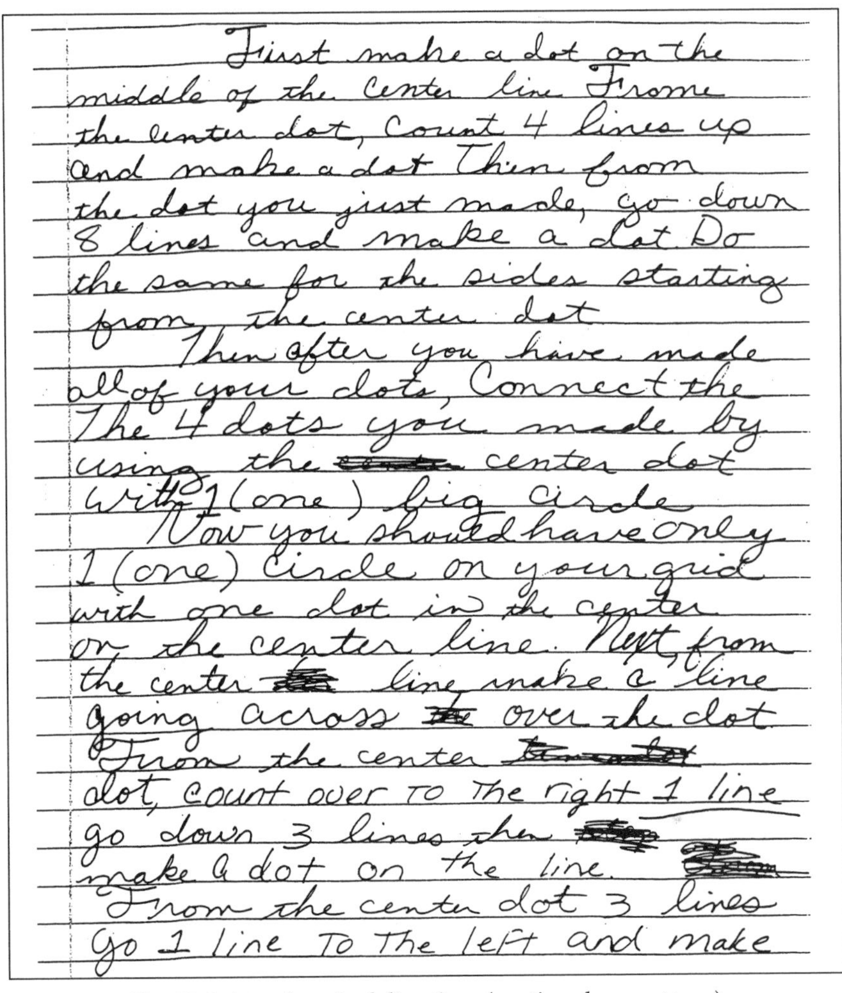

First make a dot on the
middle of the center line Frome
the center dot, Count 4 lines up
and make a dot Then from
the dot you just made, go down
8 lines and make a dot Do
the same for the sides starting
from the center dot
Then after you have made
all of your dots, Connect the
The 4 dots you made by
using the center dot
with 1 (one) big circle
Now you should have only
1 (one) Circle on your grid
with one dot in the center
on the center line. Next, from
the center line make a line
going across over the dot.
From the center
dot, count over To The right 1 line
go down 3 lines then
make a dot on the line.
From the center dot 3 lines
Go 1 line To The left and make

Fig. 20.6. Aaron's revised directions (*continued on next page*)

design on the grid that they thought the directions communicated. Figure 20.7 shows the drawing that another student made from Aaron's revised directions.

Day 3, first-hour class. Linda returned the revised directions and drawings for students to assess the extent to which their revised directions produced more-accurate drawings. This activity also led to discussions about whether the drawer had carefully followed the directions as they were written.

As teachers examine the work of Aaron and three additional students from the first-hour class, they consider the following questions:

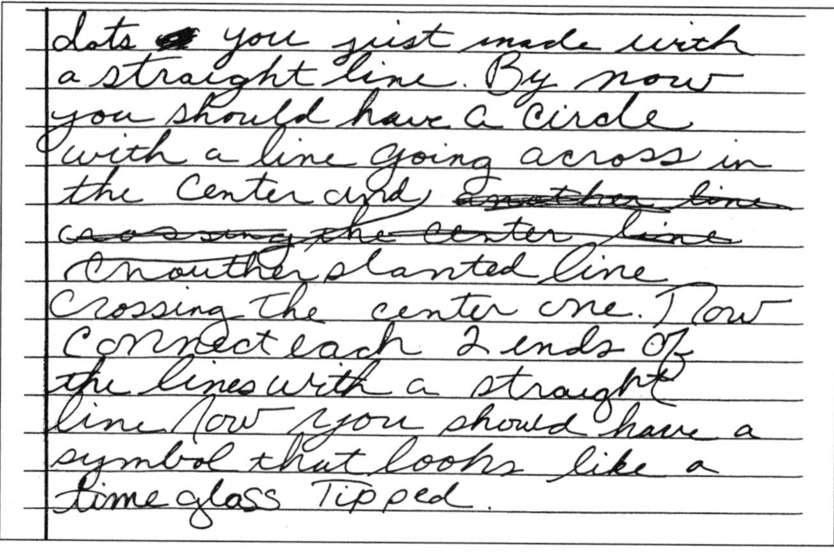

dots you jist made with
a straight line. By now
you should have a circle
with a line going across in
the Center and ~~another line~~
~~crossing the center line~~
anouther slanted line
Crossing the center one. Now
Connect each 2 ends Of
the lines with a straight
line Now you should have a
symbol that looks like a
time glass Tipped.

Fig. 20.6 (*continued*). Aaron's revised directions

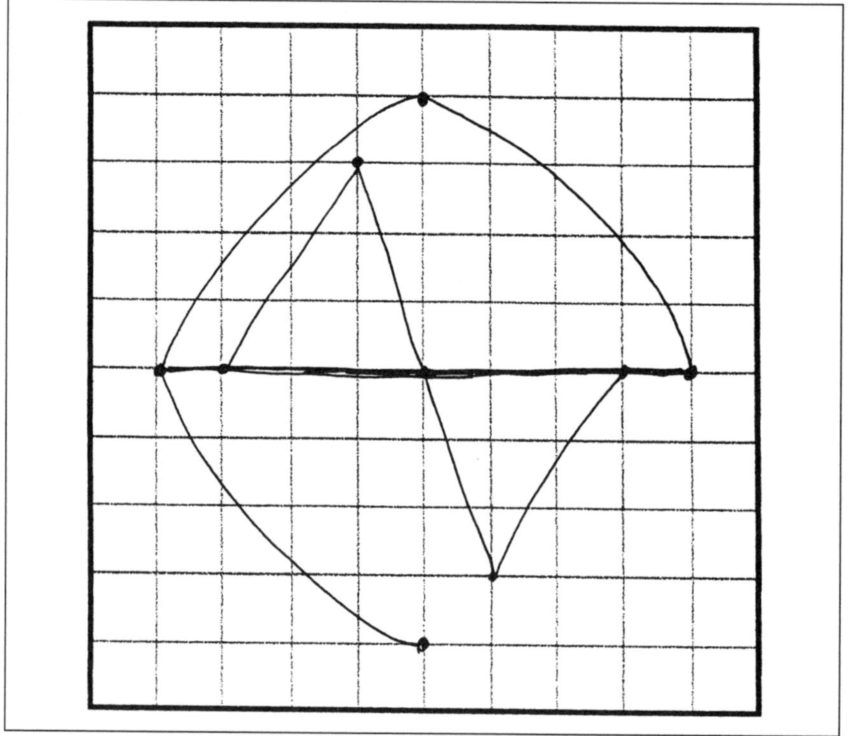

Fig. 20.7. Drawing made by a student using Aaron's revised directions

- How do you see assessment and learning both taking place?
- What ways might you have students use peer feedback on responses to other types of performance tasks?
- How and what might students learn in the process?

Teachers often remark that they use peer review and feedback in literacy but have never considered its use in mathematics. They have students exchange mathematics assignments to check answers, but the practice isn't conceived as a way of giving students feedback for the purposes of revising their work. Teachers are generally impressed with the level of attention that Aaron gave to writing his revised directions—a clear indication that he was using another student's drawings to guide him in the process. The discussion typically turns to generating ideas about how to modify the assignment so that all students have an opportunity to write a set of directions, make a design from someone else's directions, and write a set of revised directions.

Scoring What We Value

It is in scoring that we make explicit what we value in mathematical performance. In the professional development series, we deliberately postpone serious work with scoring until the last third of the sessions. Our experience is that far too often teachers equate assessment with grading. From the beginning we want to broaden teachers' stance on assessment—to view it primarily as using a variety of means to gather evidence of what students are coming to understand so they can monitor and adjust their own teaching. When questions of scoring or grading are raised in the early sessions, we acknowledge its place in a teacher's practice. But at the same time, we stress the importance of grappling with issues about students' understanding that often get superficial attention if the focus is on grading.

Using rubrics to give feedback

This activity is designed to raise issues about the kinds of feedback teachers give to students and to consider the role that rubrics can play in this regard. (We are grateful to our colleague Malcolm Swan for his conceptualization of this activity.) Teachers begin by reading the directions written by several students and respond to the following:

- How would you assess the work as a busy classroom teacher?
- Write a response to the student that gives feedback in a form that she or he would read it.

After they write feedback to the students, teachers consider the following:

- What did I value in the response? Mathematical language? Techniques? Clarity? Precision? Succinctness?

- What type of feedback did I give? Global? Specific? Encouraging? Critical?
- How would I expect the student to respond to my assessment feedback? How and what would she learn? What would I record for future reference?

Teachers generally find giving written feedback challenging. You might try writing your own feedback to Aaron on his first set of directions to see how hard this can be. These are typical comments from teachers as they struggle to give written feedback.

- "I didn't know where to start in giving feedback."
- "This takes a lot of time—how much to write and what, since I *am* a busy teacher."
- "I wanted to be encouraging, trying to keep a student's self-esteem intact."
- "I was trying to figure out what would be helpful."
- "The feedback needs to be valued by the student."

Next, teachers examine and apply two rubrics to the students' work. The aim is to get a sense of how different rubrics communicate what is valued in performance. The first scoring scheme is a task-specific, four-level, holistic rubric (fig. 20.8). The second is a task-specific, point-by-category rubric (fig. 20.9). This second rubric, with a total of 15 points available, awards points for the geometry, shape, and space content aspects of the task (GSS); for the problem-solving and reasoning aspects of the task (PSR); and for the communication aspects of the task (C).

Issues that emerge in applying rubrics

In the discussion that follows, several issues tend to emerge. When using a holistic rubric, teachers find it difficult to know how much weight to give to the various aspects of performance. Take Sasha's work, for example (fig. 20.2). There is often heated discussion about whether Sasha's work should be assessed at Level 4 or Level 3. Those who argue for Level 3 point to the fact that Sasha's system is not economic. Those who argue for Level 4 stress that even though the system is not economic, it does not result in inaccuracies in placing the design on the grid. Overall, they believe that Sasha's work meets the essential demands of the task and that the requirement of an economic system is a minor aspect of successful performance on the task. A closer look at the point rubric makes this argument even stronger. Of the 15 possible points, only 1 is awarded for an economic system.

Teachers generally believe that the holistic rubric does not tease out the geometry content aspects the way the point-by-category rubric does. They speculate that sustained use of categorized feedback (mathematical con-

T-Shirt Design Holistic Rubric

The characterization of students' responses for this task is based on these "Core Elements of Performance":

- Locate the placement of all parts of the design (circle, diameter, zigzag) on a grid
- Use an economic and systematic approach (may be an invented system)
- Give a clear set of directions

Level 4: The student's work meets the essential demands of the task.

The student gives a complete, economic system that is capable of locating all parts of the design with reasonable accuracy *and* directions are clearly stated and easy to follow.

Level 3: The student's work needs to be revised.

The student gives a system that addresses all parts of the design, which has the potential for locating points precisely, but the system is not economic, resulting in inaccuracies. Some effort may be needed to follow the directions.

Level 2: The student needs some instruction.

The student attempts a system for locating all parts of the design, but the system does not have the potential for locating points precisely, resulting in errors in placement or a lack of relationship between the whole design and its parts *or* considerable effort being needed to follow the directions.

Level 1: The student needs significant instruction.

The student uses no system for locating parts of the design and makes major errors in placement or distorts the shape of the design.

Fig. 20.8. Holistic rubric for the T-Shirt Design task

	15 points		
Step	**GSS**	**PSR**	**C**
Using a systematic approach			
• Uses effective system for locating components of the figure		**2**	
Partial credit: Some evidence of a system with potential, but results in inaccuracies or errors		(1)	
• System is economic		**1**	
The circle			
• Locates circle on the grid accurately	**2**		
Partial credit: Mentions the circle	**(1)**		
The diameter			
• Locates on the grid or circle accurately	**2**		
Partial credit: Mentions or implies the diameter	**(1)**		
The zigzag			
• Locates 4 endpoints of the segments of the zigzag			
allow 1 point for each correct endpoint	4×1		
Clarity of instructions			
• Instructions are ordered			**1**
• Instructions are clear			**2**
Partial credit: Instructions are partially clear but have essential information missing			**(1)**
• Instructions include checks for listener			**1**
Points allocated	**8**	**3**	**4**

Fig. 20.9. Point-by-category rubric for the T-Shirt Design task

tent, problem solving and reasoning, communication) to students may contribute to a different orientation to learning. Students will see that communication and reasoning are both important as well as obtaining correct answers.

Assessing Professional Development through Assessment as a Tool for Change

Do these professional development activities meet the criteria of a tool for change? Let's see. The activities are built around a specific high-quality assessment task, artifacts of students' engagement with the task, and questions for teachers' analysis and reflection. The careful design of the flow of activities extends the capabilities of experienced professional development facilitators who work with teachers in districtwide in-service programs, as well as teacher-leaders who work with colleagues in less formal arrangements in their own schools.

Most important, the routine can accommodate customization with any high-quality task and carefully chosen students' work. This is particularly useful to a facilitator who wants to use students' work collected in local classrooms. The materials are sufficiently flexible for them to be adapted easily for use in a broader program of professional development. For example, they can support the ongoing implementation of standards-based curricula by providing a classroom tool for better understanding what students are learning across a broader range of mathematical performance.

Finally, assessment can be used in a first phase of a reform agenda with a longer-term goal of implementing standards-based curricula. In this instance, teachers generally find high-quality tasks and anonymous students' work extremely interesting and want to see what their own students can do. If the curriculum-in-use is not in harmony with these kinds of tasks, it can prompt discussions about what kinds of learning experiences students would need in order to engage these kinds of problems successfully and what curriculum materials would support this.

REFERENCES

Balanced Assessment Project. *Middle Grades Assessment Package 1.* White Plains, N.Y.: Dale Seymour Publications, 1999.

Briars, Diane. "Mathematics Performance in the Pittsburgh Public Schools." Presentation at the Mathematics Assessment Resource Service Conference on Tools for Systemic Improvement, La Jolla, Calif., March 2001.

Mathematics Assessment Resource Service. *Balanced Assessment Practice Tests.* Monterey, Calif.: CTB/McGraw-Hill, 2001a.

———. *Balanced Assessment 2001 Tests.* Monterey, Calif.: CTB/McGraw-Hill, 2001b.

———. *Balanced Assessment 2002 Tests.* Monterey, Calif.: CTB/McGraw-Hill, 2002.

National Council of Teachers of Mathematics (NCTM). *Curriculum and Evaluation Standards for School Mathematics.* Reston, Va.: NCTM, 1989.

———. *Professional Teaching Standards for School Mathematics.* Reston, Va.: NCTM, 1991.

———. *Assessment Standards for School Mathematics.* Reston, Va.: NCTM, 1995.

———. *Principles and Standards for School Mathematics.* Reston, Va.: NCTM, 2000.

Ridgway, James E., Rita Crust, Hugh Burkhardt, Linda Fisher, and David Foster. *MARS Report on the 2000 Tests.* San Jose, Calif.: Mathematics Assessment Collaborative, 2000.

Senk, Sharon, and Denisse Thompson, eds. *Standards-Based School Mathematics Curricula: What Are They? What Do Students Learn?* Mahwah, N.J.: Lawrence Erlbaum Associates, 2003.

21

Teaching Mathematics with Technology
An Evidence-Based Road Map for the Journey

Karen Hollebrands

Rose Mary Zbiek

It is widely recognized that many types of technology are appropriate for high school mathematics instruction (National Council of Teachers of Mathematics [NCTM] 1989, 2000). However, technology itself is not a panacea that will remedy students' difficulties as they learn mathematics. Rather it is teachers' decisions about how, when, and where to use technology that determine whether its use will enhance or hinder students' understandings of mathematics.

Research conducted during the past two decades contains evidence of how teachers learn to use technology effectively in their classrooms. The research provides (*a*) insights into what we as teachers can expect in the process of incorporating technology effectively into mathematics teaching, (*b*) insights into why seemingly valid attempts at using technology may fail, (*c*) clues for how to address the needs and concerns of teachers, and (*d*) indications of the external factors that influence this endeavor.

To illustrate the evolution of the process of incorporating technology into teaching mathematics, a single type of technology, the Geometer's Sketchpad (Jackiw 2001), is used throughout this article. Focus on a single type of technology allows the reader to see more clearly the changes that are made as a

This article arose from the collaborative work of the authors through the Conferences on Research on Technology in the Teaching and Learning of Mathematics project (ESI-0087447) through a grant to Pennsylvania State University funded by the National Science Foundation. The ideas and opinions are those of the authors. The longer research review will appear in Zbiek, Rose Mary, and Karen Hollebrands, "A Research-Informed View of the Process of Incorporating Mathematics Technology into Classroom Practice by In-Service and Prospective Teachers," in *Handbook of Research on Technology in the Learning and Teaching of Mathematics: Syntheses and Perspectives*, edited by M. Kathleen Heid and Glendon W. Blume. Greenwich, Conn.: Information Age Publishing.

teacher progresses in an evolutionary process of incorporating technology in mathematics teaching.

A PROCESS OF INCORPORATING TECHNOLOGY INTO TEACHING

There are many things we learn to do in our lives—things at which we become proficient over time and things we eventually value as implicit parts of our daily lives. Learning to incorporate technology into the mathematics classroom is similar in many ways to learning some of these other things, such as learning how to drive an automobile. When learning to drive, a person's first behind-the-wheel experience does not take place in different cities on multiple types of roads (city streets, highways, freeways, country roads) under varied conditions (rain, ice, sleet, snow). Likewise, a first technology-using experience should not involve different types of technology (graphics calculator, dynamic program for geometry, spreadsheets) in different subjects (algebra, geometry, calculus) under varied conditions (general or honors classes of students). It is better when that first driving experience occurs in a familiar place, on a road that is easy to drive, and when conditions are most favorable. It is also better that the new driver is not alone and has the support of others. Research suggests a similar approach to the incorporation of technology in the teaching of mathematics.

Beaudin and Bowers (1997) characterized the journey of a teacher who is incorporating computer algebra systems into mathematics teaching. Their PURIA model includes five modes of use: Play, Use, Recommend, Incorporate, and Assess. Our personal experiences, as well as our review of research about the use of other forms of technology and about teachers' experiences in using two or more types of technology or in using one type of technology in two or more ways, spurred us to modify the PURIA model. We use the example of a teacher working with a dynamic environment for geometry as an example to illustrate the five modes of use.

Play to Explore the Technology

Play happens early in a teacher's use of a new technology or in the teacher's use of a familiar technology for a distinctly new purpose. It is usually a brief period. As in the subsequent modes, teachers may experience Play in the context of a workshop or class, and it may occur with some degree of guidance. Although there is mathematics involved in Play, the technology is not used for the purpose of exploring a mathematical topic, and the use is not necessarily related to how the teacher might use technology in the classroom. Play is a mode during which the teacher becomes familiar with some of the capabilities of the technology. For example, after being told by a workshop leader that the "buttons" on the left side of the Sketchpad (Jackiw 2001) screen can be used to place objects on the screen, one teacher began placing and labeling objects, creating the display shown in figure 21.1. There was no grand mathematics goal to her work; she was merely seeing what the technology could do.

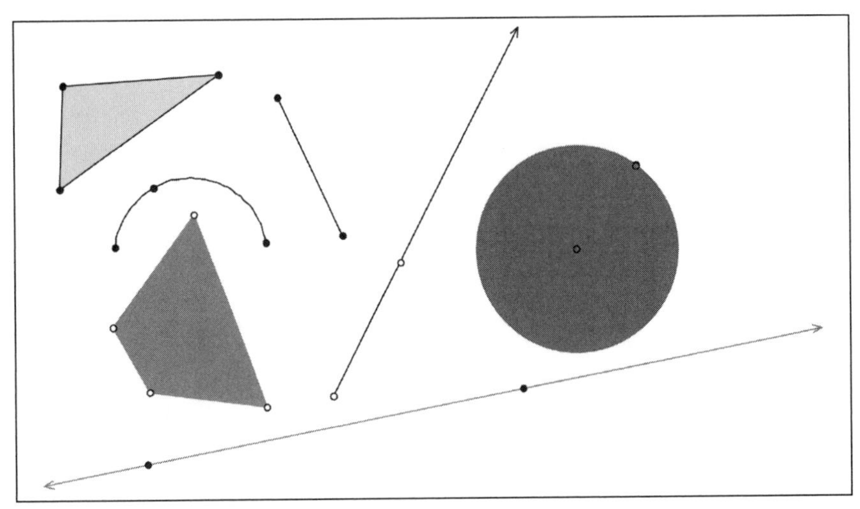

Fig. 21.1. Results of teacher Playing with Sketchpad (Jackiw 2001)

Use It for Your Goals

After initially Playing with the technology, the teacher then tries to Use the technology for a personal mathematical purpose. This Use could be influenced by reading the manual, attending a workshop, or using a tutorial that accompanies the technology. The technology might then be used to generate answers to homework problems assigned to students and solvable without the use of technology (as in fig. 21.2). Teachers may also use the technology to do mathematics for less mathematical purposes. With a graphing calculator, this might include creating drawings or screen dumps for Web sites, handouts, correspondence, or tests. Operating in the Use mode, the teacher engages the technology as a personal mathematical tool but is not yet using it in teaching.

Recommend or Regard in Limited Use

In the Recommend mode, teachers suggest the technology as a tool for use to a few students. The Recommend mode could include using the technology with a student who comes for help with difficult material or after missing a class. For example, a student may be questioning whether the altitude to the base of an

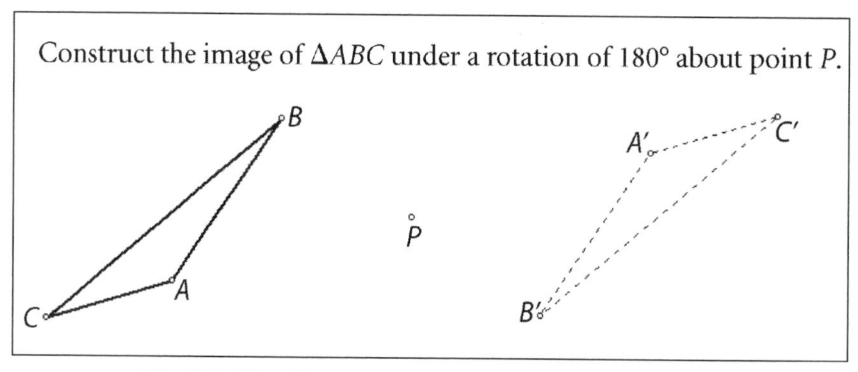

Fig. 21.2. Use of generating answer to homework problem

isosceles triangle bisects the base regardless of the size or placement of the isosceles triangle. Using Sketchpad (Jackiw 2001), the teacher may generate an isosceles triangle (ΔABC), create the altitude to the base ($\overline{CM}$, M on $\overline{MB}$), measure the base segments ($\overline{AM}$ and $\overline{MB}$) and then drag point A to show the student a variety of shapes and positions as shown statically in figure 21.3.

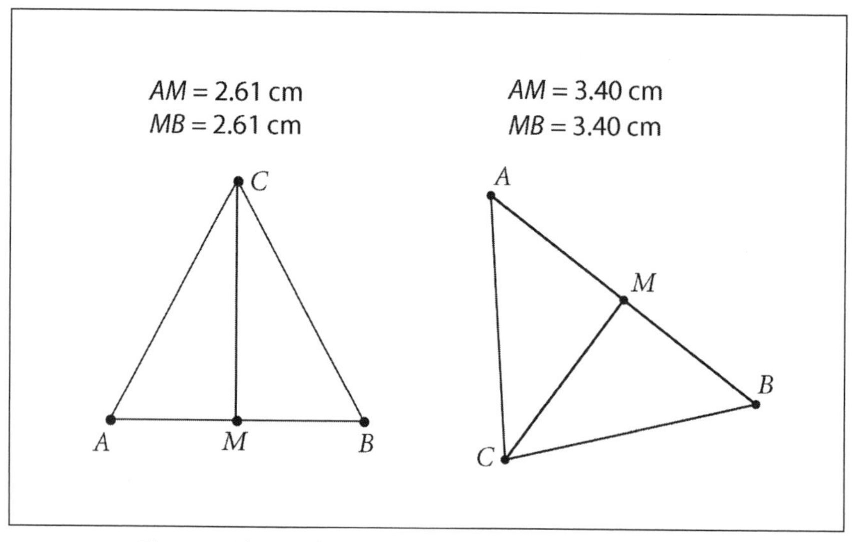

Fig. 21.3. The results of the recommendation of the teacher

The Recommend mode of use also includes watching while a student employs the technology on his or her own, in class or outside class. The Recommend mode similarly arises as two or more teachers use the technology to test a mathematical supposition in the teachers' workroom.

The Recommend mode may be an important mode not because of what it contributes to students' learning, but rather because it provides the (perhaps skeptical) teacher with an idea of how the technology fits into a mathematical setting involving other people. The teacher may reflect on these experiences and consider their implications for the classroom. This can be a bridge for the teacher to cross from thinking about the technology as "my tool" to seeing it as "their tool"—a change of thinking that prepares the way for actual classroom use.

Incorporate into a Classroom Setting

The Incorporate mode has many layers and may be a period in which a teacher works for a substantial time. Classroom experimentation and reflection on those attempts characterize the teacher's work during this mode. The teacher may start on a very small scale with a demonstration, such as one to illustrate the Pythagorean theorem as implied in figure 21.4. The teacher may then move to giving students direct contact to the technology. This could be through the use of a prepared activity from ancillary material or from an activity book designed to go with Sketchpad (Jackiw 2001). After trying a lesson or two, the teacher may refine the lesson, use it again, and eventually incorporate the technology use into other lessons. The subsequent technology-using activities may become less scripted or less prescribed by others, as when the teacher capitalizes on an impromptu question in class that the students then pursue with technology.

Later in the Incorporate mode of use, the teacher becomes more in control of the activities and more flexible in classroom use of the technology. The

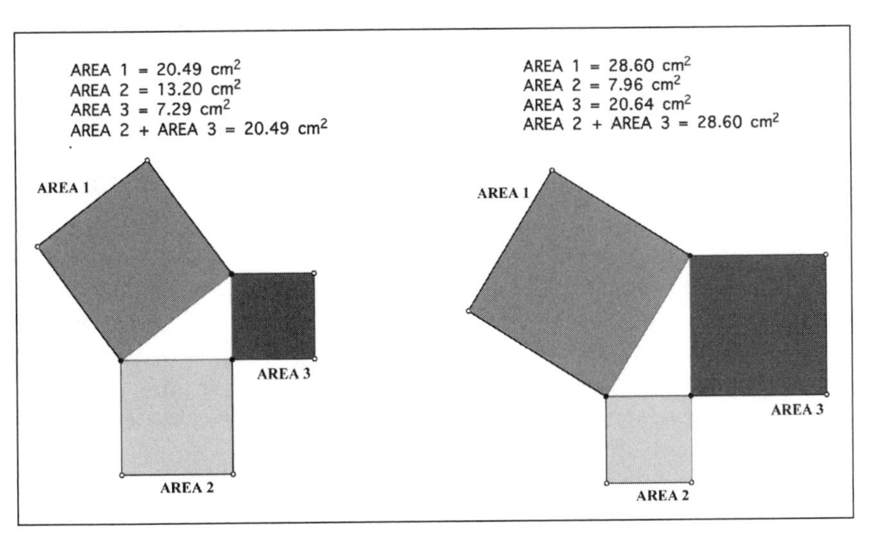

Fig. 21.4. Static sketches from a Pythagorean relationship demonstration

presence of the technology is assumed, and the teacher has comfort and confidence in using the technology. The teacher has moved from thinking about personal or students' use of the technology to using the technology to enhance mathematics lessons. The teacher's evolving expertise is now evident in the teacher's classroom practice.

With increased attention to students' developing understandings and reflection on technology use, the teacher in the Incorporate mode becomes more purposeful in choosing among electronic technology, manipulatives, and abstract approaches. For example, the topic of a lesson may be the area of a parallelogram. The teacher whose students do not understand area as covering may choose to use manipulatives to emphasize area as covering. For a class comfortable with the meaning of area and the formulas for triangles and rectangles, the teacher may choose an abstract approach and have students derive the formula for the area of a parallelogram. If students need further experience to understand length and height as variables and to see parallelogram area as a function of these two variables, the teacher may challenge students to use the dynamic capabilities of Sketchpad (Jackiw 2001) to generate many different parallelograms, each with the same area.

The Incorporate mode builds slowly until we have a technology-intensive setting in which the teacher uses the technology regularly and has curriculum and materials that draw appropriately on the technology available. Continuous reflection on classroom events underlies the teacher's evolving ability to capitalize on technology. Jumping too quickly to this technology-intensive setting may be one reason why some technology innovations fail. Concerned teachers with inadequate experiences in the prior modes may be underprepared technologically (with too little Play and Use) and unaware of how people work with technology (from too little Recommend and early Incorporate). They may become easily disheartened by the difficulties they experience.

Assess Incorporation on the Basis of Students' Learning

Incorporation alone is not the final goal. Rather, the Incorporate mode should blend naturally into the teacher's attention to students' learning. Teachers in the Assess mode go beyond concentrating on technology and lesson development as they focus on what students are learning and thinking with the technology. This locus of attention may begin to arise during the Incorporate mode, especially for experienced teachers already accustomed to concentrating on outcomes of students' learning. At the pinnacle of the Assess mode, the teacher's attention to students' learning is central to the teacher's work. Assess is where the teacher focuses on the learning and uses the quality of the students' experiences to alter or refine lessons and the associated technology use.

The Iteration and Blending of Modes

As suggested by the blending of Incorporate and Assess in the previous paragraphs, we do not view movement through the PURIA model as a linear progression. We also noted that some modes of use may be brief and that modes may be revisited. For example, consider the teacher who uses Sketchpad (Jackiw 2001) in geometry classes and reaches the Assess stance as outlined in the previous section. Suppose that the teacher learns from a colleague that Sketchpad (Jackiw 2001) can be used for Cartesian graphing. The teacher may then return to the Play mode to try the graphing capability and perhaps generate a parabolic graph to use in a handout for algebra students, thus operating in the Use mode. The teacher may come across the idea of slider graphs (see NCTM 2000, p. 338 and *E-example* 7.5, Web site standards.nctm.org/document/eexamples/chap7/7.5/index.htm) and may try that activity with some students in a mathematics club—indicative of Recommend work. The teacher may then use slider graphs for other types of functions and eventually make these a standard part of the algebra course, having moved to Incorporate. Attention to students' learning and continued refinement and use of the graphing potential, perhaps through an action research project, would have the teacher operating now in the Assess mode with a different feature of the same Sketchpad (Jackiw 2001) software.

Given a generally familiar technology environment as in this slider graph example, the teacher's second pass through the PURIA modes may be much more rapid than the first pass. However, research does suggest that the modes for the incorporation of a considerably different second form of technology may take longer, may vary in time-per-mode, and may have more or less productive results over a similar time period.

The Classroom Experience from the Teacher's Perspective

As the teacher passes through the various PURIA modes and begins to use technology with students, she may find it helpful to think about different roles to assume and to consider these roles in reflecting on classroom events. Research conducted in technology-enriched high school mathematics classrooms (Farrell 1996; Heid, Sheets, and Matras 1990) identified a variety of teachers' responsibilities, challenges, and roles. Some roles arise naturally and frequently for the technology-using teacher: Technical Assistant, Counselor, and Collaborator.

- Technical Assistant: The teacher helps students with hardware and software difficulties.
- Counselor: The teacher is familiar with the problem and is able to advise and assist students when they ask for the teacher's input. This

role includes playing the devil's advocate as well as providing encouragement or serving as a stimulator or diagnostician.

- Collaborator: The teacher is initially unfamiliar with both the problem and the solution and therefore is a true participant in mathematical learning.

Although these and other roles also arise in mathematics classrooms where technology is not present, these roles are useful tools for thinking about technology-based lessons and how good teaching with electronic technology parallels good teaching without it.

As an example, consider the instance of a teacher who is working in the Incorporate mode and is teaching a lesson with the goal of students using their understanding of families of functions as well as curve-fitting technology to model a skateboard situation (Zbiek and Heid 1990). Using a physical set-up such as the golf ball and ramp shown in figure 21.5, students collected the data shown in figure 21.6 to model the time it took a golf ball to roll down a ramp as a function of the height of the raised end of the ramp.

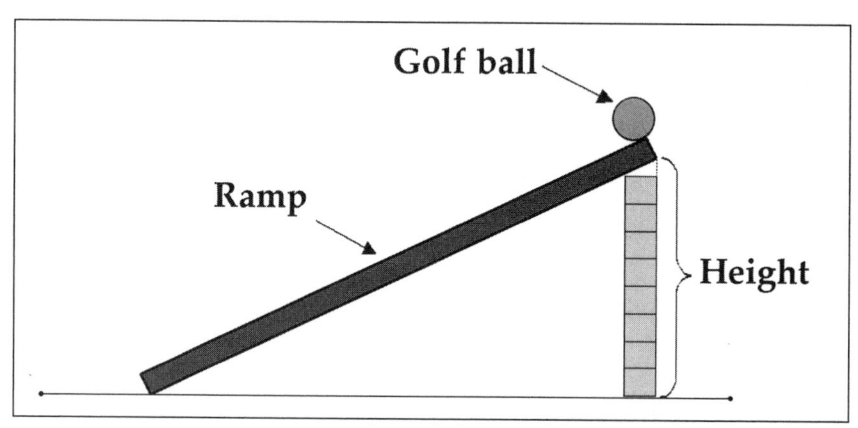

Fig. 21.5. The golf ball and ramp model for the skateboard problem

L1	L2	L3 3
2	2.445	
4	1.725	
6	1.355	
7	1.335	
------	------	

L3(1)=

L1	L2	L3 3
8	1.22	
9	1.11	
11	.96	
5	1.48	
------	------	

L3(1)=

Fig. 21.6. Golf ball and ramp data

The teacher chose to use physical materials rather than a simulation for the data collection. She thought the experience of seeing the golf ball roll at different speeds would help students connect this with their real-life experiences. Further, she believed the discussion of the models would be richer if students had noise in the data that they could interpret as the results of their own actions during data collection. Knowing the data collection set-up was crucial to the success of the lesson, the teacher chose to begin the class as a Technical Assistant helping the students arrange the physical materials. The teacher facilitated students' independence from her in a Technical Assistant role by asking students to suggest what to do next and then allowing them to test their plans.

As the students then assumed leadership roles in collecting their data, the teacher moved into a Counselor role. She challenged them to think about their work and the implications of their data collection. For example, she asked them to compare different rolls of the golf ball to bring their attention to the errors they were introducing into their data collection by, for instance, releasing the ball from different positions at the end of the ramp. She anticipated the whole-group discussion of the function models and therefore asked questions during the data collection to draw students' attention to function characteristics. As an example, after students collected data for heights of 2, 4, and 6 units, she asked them to predict the time for 7 units and to explain their rationale. She did this to draw out students' intuitions about the always-negative rate of change of time with respect to height. In reflecting on the lesson, the teacher noted this point became crucial in the students' later discussion of the functions that best modeled the skateboard situation.

At the teacher's direction and on completing their data collection, the students went to their electronic technology and most of them began Playing with the various menus and options. The teacher asked for their attention and intentionally assumed a Technical Assistant role as she told them what buttons to push and gave them ample time to master how to enter a data pair and to fit a linear, quadratic, cubic, quartic, or exponential function. This role made sense to her—and to her students—now as well as when they were assembling the physical setting to collect data. The teacher's direct guidance with the technology was an efficient way to introduce the technology skills needed by the students for the mathematical work at that moment. When the teacher observed that students were able to produce the fitted functions, she moved as planned into a Counselor role. Now she was challenging the students to think beyond merely pushing buttons and generating coefficients to consider how function characteristics such as monotonicity and domain could be used to argue for or against different models.

However, in planning the lesson, the teacher had neglected to think about how students may themselves initiate Play with the technology and how this Play can lead students to Using the technology for their own mathematical purposes. One three-person group of students had been trying various keys until one of the three students discovered he could superimpose the graph of a function of his choice over the scatter plot. The trio then began to make up a function rule to match the data points. At this point, the teacher decided to abandon her Counselor role and become a Collaborator. She and the students worked together on the novel problem of determining a function that would match the data points well. With much discussion of function characteristics such as domain, range, intercepts, and slope, the group arrived at its preferred model,

$$f(x) = \frac{4}{x^{0.6}},$$

as shown in figure 21.7.

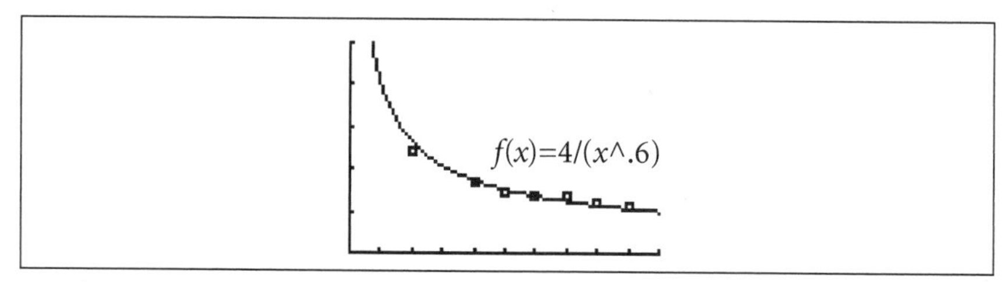

Fig. 21.7. Students' preferred model of the skateboard data

The presence of the electronic technology in this lesson afforded students an opportunity to pose a novel problem and to enter a different area of mathematics. As the students and teacher made their way along this less familiar route, their conjectures led to the development of a new solution that yielded interesting mathematical connections. The ability of technology to allow students to explore roads unfamiliar to the teacher or not part of the journey the teacher had planned required the teacher to draw on her knowledge of more familiar roads and to stay focused on the goal of the lesson. The teacher decided to pursue the students' task, as a Collaborator, on the basis of her observation that students were both highly engaged in this alternative task and still achieving the lesson goal of using function characteristics to evaluate a model. On the basis of her assessment, the teacher pursued tasks to build on what the students discovered in ways that contributed to students' developing the desired understandings of the mathematics. The new task arose spontaneously and was not found in mathematics textbooks or other sources.

EXTERNAL FACTORS

Especially during Play and Use it is helpful to have experienced support staff to assist with the technical details. The technology support you might need may not be available at your own school but might come from a person from another school or from a technology-using virtual community.

During Play, Use, and early Incorporate, access to technology is an issue. Although classrooms may not be equipped with the latest technological tools, it is important to start with something. It could be possible to use a borrowed computer or calculator for demonstrations. Materials needed to capitalize on the power of technology tools may be found through new mathematics curriculum projects and in professional journals, as illustrated with the skateboard situation. If students develop deeper understandings of the mathematical concepts when they first learn them, then there may be less need for review and reteaching. Attending workshops or taking formal classes may facilitate movement from Play through Incorporate. Informal learning and working with colleagues also helps.

CONCLUSION

Changing our classroom practice is a daunting journey. However, knowing what to expect may ease a teacher's worries. Knowing that the extent to which a teacher uses technology may reflect different modes of use (Play, Use, Recommend, Incorporate, Assess) allows teachers to create a plan with sufficient time to engage in each type of use before eventually incorporating technology into the classroom. Once a teacher does begin to use technology into the classroom, it is helpful to be aware that the teacher may assume a variety of roles (e.g., Technical Assistant, Counselor, Collaborator). It is important to keep in mind that external factors will influence, but need not hinder, the adoption of any teaching innovation. We hope this advice will assist teachers in incorporating technology in their teaching and bring closer the worthy destination of students' deep understanding.

REFERENCES

Beaudin, Michel, and David Bowers. "Logistics for Facilitating CAS Instruction." In *The State of Computer Algebra in Mathematics Education,* edited by John Berry, John Monaghan, Manfred Kronfellner, and Bernhard Kutzler, pp. 126–35. Bolton, Lancashire, England: Chartwell-York, 1997.

Farrell, Ann. "Roles and Behaviors in Technology-Integrated Precalculus Classrooms." *Journal of Mathematical Behavior* 15 (March 1996): 35–53.

Heid, M. Kathleen, Charlene Sheets, and Mary Ann Matras. "Computer-Enhanced Algebra: New Roles and Challenges for Teachers and Students." In *Teaching and Learning Mathematics in the 1990s,* 1990 Yearbook of the National Council of Teachers of Mathematics (NCTM), edited by Thomas J. Cooney, pp. 194–204. Reston, Va.: NCTM, 1990.

Jackiw, Nick. The Geometer's Sketchpad, Version 4.0. (Computer software.) Emeryville, Calif.: KCP Technologies, Inc., 2001.

National Council of Teachers of Mathematics (NCTM). *Curriculum and Evaluation Standards for School Mathematics.* Reston, Va.: NCTM, 1989.

———. *Principles and Standards for School Mathematics.* Reston, Va.: NCTM, 2000.

Zbiek, Rose Mary, and M. Kathleen Heid. "The Skateboard Experiment: Math Modeling for Beginning Algebra." *Computing Teacher* 18 (October 1990): 32–36.

22

Teachers Teaching Teachers
General Issues Section

Janet Warfield
Cheryl A. Lubinski

THE vision of mathematics teaching and learning embodied in the NCTM's *Principles and Standards for School Mathematics* (2000) and in other reform documents (NCTM 1989, 1991; National Research Council 2001) has led to the development of new teaching materials that require new ways of teaching. Encouraging teachers to help one another learn to teach in new ways has promise as a way of developing learning communities in which teachers continue to examine and improve their practice. We were interested in learning about the challenges of such an endeavor by helping teachers become teacher-leaders. Our project supported a group of elementary school teachers in their initial attempts to become teacher-leaders. We would like to share their story and our learning.

The "Teachers Teaching Teachers" Project

We worked with a group of six teachers of grades K–3 from Washington Elementary School as they planned for and implemented two professional development sessions for twenty-five teachers at nearby Jefferson Elementary School (names of schools have been changed). The Washington teachers had all participated in several Cognitively Guided Instruction (CGI) workshops and had used the principles of CGI in their own teaching. The intent of the professional development sessions these teachers conducted was to present Jefferson teachers with an introduction to CGI. Our research focused on the thinking and learning of the Washington teachers throughout the project. Our questions were as follows: On what did the Washington teachers focus as they planned for, implemented, and reflected on the pro-

The work described in this article was supported with funds from the Illinois State Board of Education (ISBE). Any opinions expressed herein are those of the authors and do not necessarily reflect the views of the ISBE.

fessional development sessions? What did the Washington teachers learn during the project?

COGNITIVELY GUIDED INSTRUCTION

Our project was grounded in the theoretical framework of CGI, which has proved to be a particularly effective professional development project (Carpenter and Fennema 1988; National Research Council 2001). Teachers who are involved in CGI are given opportunities to learn about research on children's mathematical thinking and to consider how that information can help them learn about their own students' thinking and how they can use that thinking to make more-informed instructional decisions.

The research-based information that teachers learn about at CGI workshops is organized into a framework with two main components. First, there are several types of word problems that young children are able to solve without explicit instruction. These problems are categorized according to the action or relationship in the problem as well as the location of the unknown. Consider, for example, the following problem: *Allison had 6 pennies. Jae-Meen gave her some more pennies. Now she has 13 pennies. How many pennies did Jae-Meen give Allison?* The action in the problem is a joining action; Allison got more pennies to join with the ones she already had. The unknown in the problem is the number by which the number of Allison's pennies changed. Therefore, the problem is classified a join change-unknown problem. The entire categorization scheme consists of eleven types of addition and subtraction problems and three types of multiplication and division problems (Carpenter et al. 1999).

Second, there are different strategies that children use to solve these problems. These strategies fall into three main categories: direct-modeling strategies, counting strategies, and fact strategies. *Direct modeling* involves using counters to represent all the objects in the problem to model the action or relationship in the problem. For the problem just mentioned, this entails counting out six counters, counting out more counters until there are thirteen altogether, and counting the second set to get the answer of 7. *Counting strategies* also involve following the order of the action in the problem. However, a child using a counting strategy does not represent all the objects in the problem. For this problem, a child would say "six" and then count on from seven to thirteen, extending one finger on each count until 13 was reached. The answer would be the number of fingers extended. *Fact strategies* are of two types. For this problem, a child using a *derived fact* might say, "The answer is seven, because six and six are twelve, so six and seven are thirteen." A *recalled-fact strategy* entails knowing the fact called for in the problem; for this problem, this means knowing that six plus seven equals thirteen (Carpenter et al. 1999).

At CGI professional development workshops, teachers watch videotapes of children solving word problems and discuss the problems and the strategies used to solve them. Teachers also discuss the relationships between problem types and strategies and ways to encourage children to move from using concrete direct-modeling strategies to using more-abstract counting and fact strategies. The teachers are supplied with written materials describing analyses of children's thinking, and they are encouraged to assess the validity of the analyses with their own students.

Although teachers who attend CGI workshops are not given prescriptions for instruction, there are several principles inherent in the workshops (Fennema et al. 1996, p. 407):

> (*a*) children can learn important mathematical ideas when they have opportunities to engage in solving a variety of problems; (*b*) individuals and groups of children will solve problems in a variety of ways; (*c*) children should have many opportunities to talk or write about how they solved problems; (*d*) teachers should elicit children's thinking; and (*e*) teachers should consider what children know and understand when they make decisions about mathematics instruction.

Research conducted on teachers who have attended CGI workshops and on the learning of those teachers' students has consistently shown that these teachers make changes in both their beliefs and their teaching practice and that their students outperform students in control groups on both problem solving and computation (Carpenter and Fennema 1988). A longitudinal study of CGI teachers indicated that the changes in teachers' beliefs and practice are sustained over time and that many CGI teachers exhibit generative change in that they continue to learn and grow, even when they are no longer participating in the workshops (Fennema et al. 1996).

THE CONTEXT

Washington Elementary School is a grades K–3 school in a small midwestern town. The town is about ten miles from a mid-sized city where a state university is located. The principal and some of the teachers at Washington were interested in creating a model school in which all the teachers had attended CGI workshops and used the CGI philosophy in teaching mathematics. The intent was that Washington would serve as a exemplar of effective mathematics teaching for teachers from other schools and as a clinical site for preservice teachers from the nearby university. If this goal were to be realized, the teachers at Washington would need to be better prepared to work as mathematics educators for newly hired teachers in the school, visiting teachers from other schools, and preservice teachers.

The Teachers

The six Washington teachers had all participated in several CGI workshops and grant-work activities for more than seven years. These professional opportunities had all been provided by one of the authors of this article. Since becoming involved with the CGI philosophy, they had worked with university students who learned about CGI in their mathematics methods courses. Thus, these teachers were continually thinking and learning about CGI. To further their understanding and to work toward achieving the goal of creating a model CGI school at Washington Elementary School, six of the Washington teachers agreed to conduct two two-hour CGI workshops for the principal and twenty-five teachers at Jefferson Elementary School, a grades K–5 school. Two of the six Washington teachers taught all-day kindergarten, two taught first grade, one taught second grade, and one taught third grade.

The Process

Two two-hour workshops were offered to the teachers at Jefferson Elementary School, one in early December and the other in mid-February. The purposes of the workshops were to introduce the Jefferson teachers to CGI and to offer the Washington teachers opportunities to learn to teach others about CGI. Prior to each workshop, we met for two two-hour sessions with the Washington teachers to plan for the workshops. In addition, we met again for one hour early in March to discuss the February workshop. Between meetings, we corresponded with the Washington teachers by e-mail. We also attended the workshops at Jefferson Elementary School.

Just as CGI teachers do not tell their children how to solve problems, we did not tell the Washington teachers what to do during the workshops. We did, however, plan our meetings with them in such a way as to help them think about how they would structure those workshops. At each meeting, we took extensive field notes on what the Washington teachers did and said and used those notes to help us plan for subsequent meetings. That is, in much the same way that CGI teachers use what they learn about their children's mathematical thinking to make instructional decisions, we used what we learned about the teachers' thinking to make decisions about how we would work with them.

OUTCOMES

According to Loucks-Horsley et al. (1998), teachers who are selected to become professional developers should know, have used, and be enthusiastic about the program they are to teach. We thought that the Washington teachers had the necessary skills and knowledge to conduct CGI workshops because of their experiences as in-service teachers in such workshops and as classroom teachers using CGI, as well as their experience working with pre-service teachers. Therefore, we anticipated that our role in planning and implementing the Jefferson workshops would be minimal. Although we did not participate in implementing the workshops, we found that we played a greater role than anticipated in planning those workshops.

SESSION 1: NOVEMBER 13, 2001

We met with the Washington teachers in mid-November to begin planning the first workshop at Jefferson. Before that meeting, we saw our role as simply supportive. The teachers would plan the workshop, and we would be there as they planned in the event they had questions that we might be able to answer. Two themes arose from the teachers' discussion at that meeting—themes that led us to rethink our roles in the planning process. First, although the teachers believed that children do not learn mathematics by being shown procedures, they believed that "telling" the Jefferson teachers about CGI was an appropriate way to conduct the workshops. Second, there was a strong belief, primarily on the part of the third-grade teacher, that teachers of the upper elementary grades need to do more direct instruction than teachers of lower grades do.

Telling about CGI

The discussion began with the teachers brainstorming about what they would do in the two-hour workshop. Early in the discussion, a teacher suggested that they explain the problem types. Others went on to say that in addition to teaching the problem types, they should teach the strategies children use for those problems, which problems are more difficult, and how to get children to explain their thinking. They should also include information on what CGI is, information about how to begin to use CGI, and information about how to use children's literature to create word problems. Underlying most of this discussion was the notion that they would "tell" or "explain" all this information to the Jefferson teachers.

Third Grade Is Different

Another thread that ran through the discussion was raised by the third-grade teacher. She contended that in third grade, teachers need to prepare their students for the state exam that is given toward the end of third grade and to ensure that their children meet the expectations of the fourth-grade teachers. For her, this meant that children needed to know standard algorithms. When challenged by the other teachers, she replied that the children not only needed to know the algorithms but also needed to know other mathematics content as laid out in the state's standards, including geometry, telling time, money, and measurement. Since there was so much she had to teach, it wasn't possible to spend much time on CGI. She thought it was important to make the Jefferson teachers aware of this.

E-MAIL MESSAGES: NOVEMBER 15, 2001

We left the first session at Washington perplexed and also concerned about what we had learned from listening to the teachers. We were pleased that they were able to pick out principal components of CGI. However, we were concerned for several reasons. First, instead of seeing that the purpose of the two workshops was to provide the Jefferson teachers a short introduction to CGI, they were thinking about teaching what is usually taught in a forty-hour workshop in two hours. Second, they were going to teach about CGI by "telling" the Jefferson teachers what they needed to know. And third, the third-grade teacher seemed determined to include reasons why CGI doesn't work in the upper elementary grades. After reflecting on the session and talking at length about it, we decided to send an e-mail message to the Washington teachers. We hoped that through this e-mail, we could encourage the teachers to reflect on the previous day's discussion. Accordingly, we wrote the following:

> We have been thinking and talking about our meeting yesterday. We would like to create a dialogue among all of us before our next meeting, which is on November 27th. Some questions we would like to begin with are: How long did it take you to understand what CGI is and to learn the problem types and strategies? What helped you to understand CGI? Also, we believe that since we only have four hours with the Jefferson teachers that the focus should be only on CGI AT THIS TIME, and not on testing, textbooks, or state standards. These are complex issues that are not easily resolved, and four hours is clearly not enough time to begin to discuss them. However, they are important issues and are not to be ignored in the long run.

Our hope was that by posing these questions, we might help the teachers rethink the ideas that they could teach everything there is to know about CGI in two hours and that they should tell the Jefferson teachers about CGI. We also hoped that by asking them to focus only on CGI, we might defuse the third-grade teacher's insistence that the Jefferson teachers should be told that CGI doesn't work well in the upper elementary grades.

Each of the six teachers replied to the e-mail. The answers of five of the teachers were similar in that they said they were still learning about CGI and that what had helped them learn was watching their children solve problems. One first-grade teacher wrote:

> A loooongg time! At first (1994) I just thought it was problem solving, … I remember using it on "Fridays" or whenever. Then I took more classes on it, and

> learned that it is so much more than just problem solving. The classes, discussions, and videos of children doing CGI were good (especially when I first started), but I think what really helps me understand CGI is to lead my children into the problem solving, watch them solve the problems, and talk to my colleagues about the problem solving that the children are doing. I think I keep growing in CGI, but will never feel like I've mastered it.

The third-grade teacher, who had not seemed to understand the principles underlying CGI at the previous day's meeting, wrote that she understood CGI:

> It took me several weeks to understand the basic CGI concept and problem types. I am still learning strategies as I work on problems at my level. Two things helped me understand CGI. One was to change my way of thinking and teaching. I have learned to accept more than one strategy to solve a problem. The other was for me to solve problems on my own level that were just as hard for me as what I give my students.

She did not mention learning from her students, as all the other teachers did.

SESSION 2: NOVEMBER 27, 2001

From what we learned at the first session and read in the e-mail messages, we decided that we needed to increase our role in the second session. We planned an agenda for the session and questioned the teachers about their ideas for the workshop. We began the meeting by asking the teachers to spend five minutes thinking silently about the question: What is CGI from the teacher's point of view? After five minutes, the teachers shared their responses to the question. CGI, they said, focuses on (1) understanding, (2) teachers allowing students to solve problems in their own ways, and (3) the importance of communication among children. They noted that CGI is time-consuming and that it can be cross-disciplinary. Their answers indicated that they saw CGI as a procedure to be followed (let students solve problems and talk about how they solved them) rather than a theoretical framework that links what is known about how children learn mathematics and ways that information can be used to make decisions about how to help specific children learn.

We next asked the teachers to begin discussing a plan for the first two-hour session at Jefferson. Throughout the discussion, we questioned the teachers about how the activities they were planning would contribute to the Jefferson teachers' understanding of CGI. The teachers began by saying that there would need to be introductions. They suggested asking each Jefferson teacher to introduce herself (all the teachers were women) and saying what she expected from the workshop and why she was there. We asked how long that would take and how the answers to those questions would help them work with the teachers. The teachers then decided that they would introduce themselves but would not have the Jefferson teachers introduce themselves. They decided to follow the introductions with a **KWL**, a technique that includes asking at the beginning of a class what they **K**now about the topic to be discussed, what they **W**ant to know about it, and then at the end of class asking them what they have **L**earned.

Next, they planned to show an introductory video on CGI. There was discussion about how much and which segments of the video to share. It was decided to show the parts of the video in which children's problem-solving strategies are discussed. Interestingly, the video was to be used to "tell" the Jefferson teachers about children's strategies rather than have the Washington teachers do it themselves.

The teachers also made the decision to focus the rest of the workshop on a problem solved by a third-grade class in the video. The problem was the following:

> Thirty-six children are in line to ride the roller coaster. The roller coaster has 10 cars. Each car holds 4 children. How many children can sit 3 to a car, and how many have to sit 4 to a car?

The teachers planned to stop the video after the problem was read and ask the teachers to solve it and to share their strategies with one another before showing the video of the children solving the problem. We asked what mathematics could be taught using this problem and which of the state standards were addressed by the problem. There was discussion on how to talk with the Jefferson teachers about the mathematics in the problem.

Next, a first-grade teacher said that they (the Washington teachers) should give the Jefferson teachers an assignment to do before the February workshop. The teachers could, she said, write problems of the types included in the CGI framework; those problems could be based on a children's book. However, she said, "We didn't [plan to] tell them the problem types [yet]." A heated discussion ensued about whether to have the teachers write their own problems or to give them problems from which to choose. All the teachers, however, agreed to have the Jefferson teachers ask their own students to solve word problems and question the children in order to understand their solution strategies. This decision on the part of the teachers was, in our opinions, a positive change from the first meeting when they talked about telling the teachers about children's strategies. Each of them except for the third-grade teacher had mentioned in her e-mail that she learned about CGI by working with her own students. Between the first and second planning sessions, perhaps prompted by our e-mail, the teachers had decided to build into their professional development plans opportunities for the Jefferson teachers to learn from their students. We halted the discussion and asked the teachers to think silently about this question: How will you help the Jefferson teachers understand what CGI is? The teachers claimed that all the activities they had planned would do that.

JEFFERSON WORKSHOP 1: DECEMBER 4, 2001

The Washington teachers conducted the workshop in essentially the form they discussed at the November 27 session. A slight modification was made: after the teachers had solved the roller-coaster problem and watched the video of children solving the problem, the third-grade teacher showed a list of topics taken from the state standards document and asked which of those topics were addressed with the roller-coaster problem. At the end of the two hours, they asked the Jefferson teachers to answer two questions in writing: What did you learn today? What questions do you still have? Their intent was to use the responses to these questions when they planned the February workshop. They closed the workshop with an assignment. The Jefferson teachers were to read some CGI articles they were given. Using those materials, they were to write word problems and try some of the problems with their children before the next meeting. They were also to try the following problem with their classes:

> Lizzie collects lizards and beetles. She has 8 creatures in her collection, and they have 36 legs. How many are lizards and how many are beetles?

E-MAIL MESSAGES: JANUARY 21, 2002

Between the December 4 workshop and our next session with the Washington teachers, we sent an e-mail message to the teachers, asking: What did you learn from the December 4 session at Jefferson? We received only two responses to the question. One of the first-grade teachers wrote:

> I was a little concerned that I overheard that the administrator didn't think that [CGI] could be implemented unless it was school-wide.... I don't think you can make a teacher change! You must certainly WANT to do this because you feel it is better for children. I think that exposure in any classroom is better than nothing. Secondly, I don't think the teachers saw much difference in their own "problem of the day" and CGI. Maybe we didn't concentrate on the importance of building on children's thinking as much as we did problems ... but it is hard to convey that in two sessions.... I don't think that assessing children's thinking is that easy. It has to be done over time and the little rascals keep changing! It varies with problem type and numbers so it really is hard to get a handle on. Especially when you are comfortable with giving a test ... the kids pass ... and you convince yourself you've done your job. My own personal life history tells me what a sham that is but it is still harder for teachers to accept something a little more intangible than that straight percentage. SO ... I think next session we might spend some more time working on how the skills build as you are doing CGI ... and the individual aspects of it [versus] whole class problems.

We were impressed with this response. The teacher's pedagogical content knowledge related to CGI went far beyond what any of the teachers had expressed at the planning meeting on November 27. She expressed the view that a large part of CGI lies in assessing individual children's understanding and finding ways to help the children build on that understanding. She made this clear by referring to questions asked by the Jefferson teachers about the differences between CGI and the "problem of the day" that they already did. She also acknowledged that a very important aspect of CGI had not been addressed in the session at Jefferson. Furthermore, the issue of the Problem of the Day compared to CGI had not been addressed in any of the classes the Washington teachers had taken. This provided evidence that their CGI experiences were generative.

SESSION 3: JANUARY 29, 2002

We began the session by asking: What did you learn at Jefferson? What did you learn about the teachers there and how they were thinking about CGI? The teachers responded that by listening to the Jefferson teachers during the workshop and by reading their written responses to the questions about what they had learned and what they still wanted to know, they learned that the Jefferson teachers had many of the same questions they themselves had when they were first introduced to CGI. These included questions about how CGI fits with the curriculum they are supposed to teach, the state test scores at Washington, where teaching computation fits in, how the Washington teachers manage their time to allow children to share strategies, how they determine grades, and whether there are problems appropriate for use in the upper elementary grades.

The teacher who wrote the e-mail above responded that the Jefferson teachers thought that CGI was like using a "problem of the day." She claimed that they missed the whole point of CGI. Interestingly, she acknowledged that they (the Washington teachers) had also missed the point when they began learning about CGI. The second-grade teacher elaborated by saying that CGI is a philosophy, not just a "problem of the day." Instead, it allows teachers to approach a concept with a child and to know about the whole child. At the November 27 meeting, we had asked the teachers to define CGI. At that meeting, none of them mentioned that CGI allows them to assess children's mathematical thinking and plan instruction based on that assessment. This idea didn't emerge until after the first workshop at Jefferson.

At this third planning session with the Washington teachers, we began a discussion to decide which of the Jefferson teachers' questions could be addressed at the next workshop. Issues about whether CGI is appropriate for older children—issues that had arisen in our first session with the Washington teachers—had also come up in the Jefferson teachers' question about problems suitable for children in the upper elementary grades. We had previously decided that we would like for the teachers to address this issue. We opened this discussion by re-creating, with the help of the teachers and our notes, the strategies that the Jefferson teachers had used for the roller-coaster problem. Some had directly modeled the problem using pictures. Others had used tables to represent their thinking about the problem. After discussing these strategies, we asked if the problem could be solved using algebra. Some of the teachers worked individually and some worked in pairs to try to figure out an algebraic way of solving the problem. After some time, a first-grade teacher arrived at a system of equations ($x + y = 10$ and $3x + 4y = 36$). She then solved the system, first by graphing and then by substitution. After the problem had been solved, we pointed out that the Jefferson teachers had asked for more difficult problems, but they had not recognized that the problem they worked on was challenging and could be solved using algebraic strategies.

It was apparent throughout this discussion that several of the Washington teachers did not have the mathematical content knowledge necessary to solve the problem algebraically. Later, as we reflected on the discussion, we wished we had approached the issue of more-difficult problems and strategies differently. One approach we could have taken would have been to ask the teachers how to represent their own solutions symbolically. Then we could have asked the teachers to explain the ideas represented by the symbols. They might have said: "The number of cars with three children plus the number of cars with four children is ten," or "the sum of four times the number of cars with four children plus three times the number of cars with three children is a way to describe the number of children altogether." This discussion could have led to generalizations about relationships or an algebraic representation of the situation. However, if it did, it would be a representation derived from the teachers' initial thinking about the problem and not from our suggestion about using algebra.

SESSION 4: FEBRUARY 5, 2002

This session was devoted to developing plans for the next workshop at Jefferson. A list of the Jefferson teachers' questions was written on the board, and the Washington teachers talked about which of those questions could be addressed during the workshop. They decided to address several of the questions. First, they planned to talk about classroom management. They wanted to ask what those teachers who listed it as an issue meant and whether their concerns were specifically related to CGI or whether classroom management was an issue regardless of how they taught mathematics. They also decided that they would answer questions about grading by sharing some rubrics they had developed and some that are used to score open-ended questions on the state test.

The main focus of the second Jefferson workshop, however, would be connected to strategies children use to solve problems. They would use an activity in which teachers would be given several word problems and asked how each would be solved using (1) a direct-modeling, (2) a counting, and (3) a facts strategy. They also decided that they would use the work the teachers brought showing how their children solved the Lizzie's Creatures problem. They would make connections among the strategies and include the use of algebra, as we had done at the planning meeting with the roller-coaster problem.

Workshop 2: February 19, 2002

This workshop was carried out as planned. Through listening to and watching the Washington teachers, we learned a great deal about their own understanding of problem types and solution strategies. We learned, for example, that when asked about the difference between the direct-modeling and the counting strategies, two of the Washington teachers were unable to articulate the difference. At the end of the session, the Washington teachers asked the Jefferson teachers to write the answers to two questions: What have you learned? and What should we do differently if we do this again?

Session 5: March 8, 2002

We met with the teachers for a debriefing about the entire project. We asked them to think about three questions: (1) What did you learn from doing the workshops at Jefferson? (2) What would you do differently, if you were to do it again? and (3) If we could have one more two-hour session with you, what topic would you like to consider?

After our first meeting with the Washington teachers, we were concerned about two ideas they expressed. One was the notion that they could tell the Jefferson teachers everything about CGI in two hours, and the other was that teaching third grade is somehow different from teaching kindergarten, first, or second grade. These two issues arose again in the final session, as other themes that we had noticed did throughout the project.

What We Learned and What We Would Do Differently

A kindergarten teacher said that what she had learned was that it was easier to have the teachers become involved in doing something (e.g., solving problems or explaining specific strategy types) than to tell them. That is, the ideas were clearer to the teachers when they were actually doing something than when the ideas were explained to them. She also commented that there was not nearly enough time. Her statements were a clear indication that she had moved in her thinking about professional development since the first meeting. At that meeting, she thought that in the first two-hour workshop they could tell the Jefferson teachers about problem types, strategies, which problems are more difficult, how to get children to explain their thinking, what CGI is, and how to begin to implement CGI.

A first-grade teacher commented that the Jefferson teachers were more open to new ideas than she had anticipated. Others disagreed with her. They thought that although some of the teachers were open, others were not, and that some teachers were worried about basic facts and skills. These teachers, they said, thought that if you used CGI, the only mathematics you did with children was problem solving. They also commented that they (the Washington teachers) tried to do too much by including the algebraic strategies. They said the Jefferson teachers were overwhelmed and a little scared and that everything should not have been introduced in such a short period of time. This was a big change from the first meeting when the teachers had discussed "telling" the Jefferson teachers everything about CGI in the first two-hour session.

The third-grade teacher was still concerned about the teachers of third, fourth, and fifth grades. She thought that they felt isolated, that the content of the workshop was not relevant to them. Her solution was that if they were to do this again, she would like to separate those teachers and reassure them that they could take it one step at a time. Interestingly, she said she would like to talk to them about how they could do both CGI and "regular math." She went on to say that this is the fourth year she has used CGI and she is still feeling her way and still learning.

If We Did a Two-Hour Session, What Would You Like to Do?

The Washington teachers' responses to the question about what they would like to do in an additional two-hour session gave us insight into their pedagogical content knowledge. A kindergarten teacher said that she would like to learn more about how to encourage children to move toward using more-advanced strategies. There was quite a lot of discussion on this issue. Several teachers indicated that they had students who were using direct modeling for all the problems. They had tried everything they could think of to encourage these children to begin using counting strategies, but nothing worked. As this discussion went on, we learned that the teachers had limited understanding of some of the strategies children use to solve problems. They talked, for example, about a problem such as this one:

> Trevor had some toy cars. For his birthday, he got 6 more toy cars. Now he has 13. How many toy cars did he start with?

They claimed that they had students who could use either direct-modeling or counting strategies for this problem. However, since both of those strategies require following the action in the problem, and since the number of cars at the beginning of the story is the unknown, the typical strategy used by a direct modeler or a counter is trial and error. It's often the only strategy they can use with understanding for this type of problem.

We also confirmed that the third-grade teacher's understanding of the underlying principles of CGI is limited. She talked about doing both CGI and "book teaching." To her, this meant allowing the children to solve a problem using their own strategies and then telling them how the textbook said to solve the problem. Although this concerned us, she talked about having made large advances in her understanding this year. She had earlier been having children share their problem-solving strategies, but this year she began having them compare their strategies and talk about how they were the same and different. However, she then stated, "I teach the strategies I have come up with myself." Thus, it was difficult for us to determine the extent to which she developed her students' own strategies.

CONCLUSIONS

Our original questions were these: On what did the Washington teachers focus as they planned for, implemented, and reflected on the professional development sessions? What did the Washington teachers learn during the project? We conclude with their responses to these questions and comments about our own learning during the project.

Foci of the Washington Teachers

The Washington teachers began the project focusing on the procedural aspects of CGI. They planned to teach about problem types and strategies and getting children to talk about their strategies for solving problems. Initially, they did not focus on the principles that are usually inherent in CGI workshops. In particular, they did not stress that teachers should elicit children's thinking and should use that thinking to make decisions about mathematics instruction. During the project, however, they became aware that they had not addressed these principles and that as a result, the Jefferson teachers saw CGI as similar to a Problem of the Day. The teachers also focused on the concerns and questions raised by the Jefferson teachers. Although the concerns and questions of teachers are always important, many of those raised were not specific to CGI and, as we suggested in our first e-mail, were not appropriate issues to address in only four hours. By the end of the project, the teachers had come to realize that their foci had not been appropriate, that they had included too much information in too little time.

Learning of the Washington Teachers

We observed that the Washington teachers increased their understanding during this project. They learned that the underlying principles of CGI were at least as important as the problem types and strategies. They recognized that the questions the Jefferson teachers raised were the same ones they had when they were first introduced to CGI. This enabled them to see more clearly the progress they themselves had made. In conjunction with this, however, they recognized their own need to learn more and, in particular, to understand more about helping their students progress in mathematical understanding.

Our Own Learning

We learned that even though the Washington teachers knew, had used, and were enthusiastic about CGI, they were not prepared to conduct effective professional development. They needed several types of support to be better prepared. They needed help standing back from their own use of CGI in their classrooms in order to reexamine the principles about children's learning and about teaching that are inherent in CGI. They also needed to revisit the details of the research-based information on children's thinking that is shared at CGI workshops. In addition, they needed assistance in thinking about other teachers as learners who construct their own knowledge about teaching.

References

Carpenter, Thomas P., and Elizabeth Fennema. "Research and Cognitively Guided Instruction." In *Integrating Research on Teaching and Learning Mathematics*, edited by Elizabeth Fennema, Thomas P. Carpenter, and Susan J. Lamon, pp. 2–16. New York: State University of New York, 1988.

Carpenter, Thomas P., Elizabeth Fennema, Megan Loef Franke, Linda Levi, and Susan B. Empson. *Children's Mathematics: Cognitively Guided Instruction.* Portsmouth, N.H.: Heinemann, 1999.

Fennema, Elizabeth, Thomas P. Carpenter, Megan Loef Franke, Linda Levi, Victoria R. Jacobs, and Susan B. Empson. "A Longitudinal Study of Learning to Use Children's Thinking in Mathematics Instruction." *Journal for Research in Mathematics Education* 27 (July 1996): 403–34.

Loucks-Horsley, Susan, Peter W. Hewson, Nancy Love, and Katherine E. Stiles. *Designing Professional Development for Teachers of Science and Mathematics.* Thousand Oaks, Calif.: Corwin Press, 1998.

National Council of Teachers of Mathematics (NCTM). *Curriculum and Evaluation Standards for School Mathematics.* Reston, Va.: NCTM, 1989.

———. *Professional Standards for Teaching Mathematics.* Reston, Va.: NCTM, 1991.

———. *Principles and Standards for School Mathematics.* Reston, Va.: NCTM, 2000.

National Research Council. *Adding It Up: Helping Children Learn Mathematics.* Edited by Jeremy Kilpatrick, Jane Swafford, and Bradford Findell. Washington, D.C.: National Academy Press, 2001.